SpringerBriefs in Geography

SpringerBriefs in Geography presents concise summaries of cutting-edge research and practical applications across the fields of physical, environmental and human geography. It publishes compact refereed monographs under the editorial supervision of an international advisory board with the aim to publish 8 to 12 weeks after acceptance. Volumes are compact, 50 to 125 pages, with a clear focus. The series covers a range of content from professional to academic such as: timely reports of state-of-the art analytical techniques, bridges between new research results, snapshots of hot and/or emerging topics, elaborated thesis, literature reviews, and in-depth case studies.

The scope of the series spans the entire field of geography, with a view to significantly advance research. The character of the series is international and multidisciplinary and will include research areas such as: GIS/cartography, remote sensing, geographical education, geospatial analysis, techniques and modeling, landscape/regional and urban planning, economic geography, housing and the built environment, and quantitative geography. Volumes in this series may analyze past, present and/or future trends, as well as their determinants and consequences. Both solicited and unsolicited manuscripts are considered for publication in this series.

SpringerBriefs in Geography will be of interest to a wide range of individuals with interests in physical, environmental and human geography as well as for researchers from allied disciplines.

Matteo Giacomelli

Ecologies of Cohesion

A Social-Ecological Perspective
On Territorial Cohesion

Matteo Giacomelli
Department of Architecture and Urban Studies
Politecnico di Milano
Milan, Italy

School of Architecture and Design
University of Camerino
Ascoli Piceno, Italy

Inabita Laboratorio Territoriale
Ripe San Ginesio, Italy

ISSN 2211-4165 ISSN 2211-4173 (electronic)
SpringerBriefs in Geography
ISBN 978-3-032-01158-9 ISBN 978-3-032-01159-6 (eBook)
https://doi.org/10.1007/978-3-032-01159-6

This Springer imprint is published by the registered company Springer Nature Switzerland AG
The registered company address is: Gewerbestrasse 11, 6330 Cham, Switzerland

If disposing of this product, please recycle the paper.

Foreword by Massimo Sargolini

The attempt to conceptualize the role of nature in the organization of the city and the territory dates back a long time and, in particular, stems from the recognition of the inadequacy of GDP as a measure of a nation's true wealth. Senator Bob Kennedy's speech in his 1968 presidential bid ended something like this: "GDP measures everything except what makes life truly worth living." That speech, which challenged powerful financial interests, became a landmark critique of GDP. From that moment on, a line had been drawn. The debate has since continued, marked by both advances and setbacks.

In 2015, the Committee for Natural Capital, responding to a 2011 World Bank report that questioned where the true wealth of nations lies, began measuring the physical and monetary value of natural assets—forests, biodiversity, rivers, seas, and ecosystems in general. Today, however, the number of those advocating for a strictly monetary valuation of nature and culture—and the services they provide—is steadily declining. What is gaining recognition instead is the heritage value of places, understood as a fundamental input for generating meaningful data and providing clearer orientation in territorial design processes.

Within this field, leading researchers have progressively broadened their focus, increasingly linking ecosystem services to territorial assessment and planning. Contributions from bioregional planning have further supported these trajectories, helping to shift away from an understanding of urban planning narrowly associated with land consumption (Magnaghi 2014). It is within this evolving discourse that Matteo Giacomelli's doctoral research (2023) is situated—work that has been further developed through various academic publications (Giacomelli et al. 2024) and is thoughtfully synthesized in this volume.

The author starts from a basic consideration that extracts from a broad survey of the demographic and socioeconomic trends of the inland areas, arguing negative dynamics and the need for structural interventions to restore meaning and hope for the future to these geographical areas. These are two considerations that may seem obvious, but are not at all, for the following reasons:

- Even in recent times (COVID period, for example), we have read interventions by star architects who prophesied a mass emigration toward inland areas, without considering their lack of general attractiveness, often reserved for some special users that Osti will define as "the new ascetics" (2006) and above all the absence of the minimal conditions for what Fabrizio Barca defines as the state of "citizenship" (2018).
- When the urban spaces of inner areas are redesigned, we often remain anchored to an overly traditionalist vision of urbanization to be carried out, continuing to imitate the territorial balances of areas already profoundly transformed by the urbanization processes of the valley and the coast, without taking into account the fragility and wealth of resources, sometimes unknown, that characterize these geographical areas, as in the case study of the Marche region, presented in this volume.

Building on these two lines of reflection, Giacomelli draws from the literature to introduce the overarching theme of Nature's Contributions to People (NCP), approaching them in their full scope and complexity. He highlights their potential to address many of the challenges and demands of our time: from the supply of essential resources for life—such as food and water—to the regulation and mitigation of impacts related to climate and ecological change. NCP also encompasses foundational ecological processes such as soil formation, nutrient cycling, and photosynthesis, as well as cultural contributions that manifest through recreation, experience, aesthetics, and spiritual significance.

At the core of this work lies the aim of identifying how nature's resources—understood through the lens of ecosystem services—can be integrated into urban and territorial planning. It offers orientations, guidelines, and pathways that can be further developed in future research, particularly with regard to applications in the field of green infrastructure.

In the wake of the School of Territorialists[1], as well as of the great interpretative and planning experiences carried out in the planning of protected areas[2], but also of treaties and conventions already shared at European level[3], this work explores the relevance of conceptualizing landscapes as socio-ecological systems. It also highlights the need to define and assess the role of social actors in shaping these landscapes. In this perspective, NCP emerges as a bridge connecting environmental dynamics with the lived realities of settled communities.

---

[1] The "School of Territorialists" refers to a multidisciplinary group of Italian scholars and practitioners, notably including Alberto Magnaghi, who promote a place-based approach to territorial planning grounded in local knowledge, identity, and the co-evolution of human and ecological systems.

[2] Refers to the working group coordinated by Roberto Gambino, which—shortly after the approval of Italy's 1991 Framework Law on protected areas—pioneered the integration of landscape values and territorial planning approaches into the management of protected areas.

[3] Refers in particular to the European Landscape Convention, adopted by the Council of Europe on 20 October 2000 and ratified by the Italian State on 9 January 2006, which formally recognizes landscape as a key dimension of public interest and a shared responsibility in spatial planning.

However, if we truly aim to translate these reflections into a new approach to applied research, it becomes necessary to initiate transdisciplinary paths from the outset. Several leading international institutions have already established research departments that bridge diverse fields of knowledge. Examples include Cornell University in Ithaca, USA, and the Ecosystem Services Lab at iDiv, part of the German Centre for Integrative Biodiversity Research, where ecologists, sociologists, and urban planners work collaboratively.

Research structures of this type become central to the great and innovative challenge, which characterizes the European debate also in the development of pilot actions in international research projects, such as Life and Horizon: on the one hand, the implementation of systemic and integrated interventions that combine different objectives, from on the other, the improvement of biodiversity, the increase in the quality of life in urban and peri-urban areas. This is a relevant point of the European Biodiversity Strategy for 2030 and the Ecological Transition Plan. In the Italian case, the interventions implemented in these specific contexts, located most of the time outside the formally recognized administrative areas of protected areas, can increase ecological-functional connectivity, reduce anthropic pressures, promote the transition toward sustainability, and reduce emissions, as well as promoting more virtuous behavior, awareness, and education on the value of biodiversity conservation.

However, it must be acknowledged that a significant gap remains in recognizing the contribution of communities. In particular, the interaction with stakeholders in collaborative planning processes for landscape transformation—and their effective access to decision-making—requires deeper exploration. In this sense, it is not only a matter of developing appropriate methods of engagement (for which specific professional skills exist, such as those of trained facilitators), but above all of acknowledging and conveying the level of knowledge that communities themselves are able to contribute. It is not uncommon for scientific approaches to overlook or marginalize the contribution of local knowledge, treating it as secondary or merely ancillary to the domain of the hard sciences. Yet, there is a compelling rationale for its inclusion—one that stems from the very multidimensionality of green infrastructures, which can serve as the structural foundation for landscape planning.

The landscape is shaped by a system of interdependent networks, each contributing in specific ways: (1) the ecological network, i.e., the system of natural resources that raise the environmental quality of the landscape; (2) the accessibility and use network, i.e., the system of routes that allow movement through places of high environmental and landscape quality; (3) the historical-cultural network, i.e., the system of constituent elements of the traditional historical landscape (built, agricultural, and natural), which contributes to the construction of the sense of identity and belonging to places; (4) the network of the agricultural fabric, i.e., the system of agricultural companies linked to agri-food production, which act on the territory, contributing to the care of the territory and the construction of the landscape; (5) the network of infrastructures and human settlements, or the anthropic system of settlements and connecting infrastructures; (6) the social network, or the system of relationships between the community that lives in a place and its territory, which is expressed in the knowledge, use, and valorization of local resources.

Human action contaminates, directs or, in any case, interferes with each of these networks. It is only this system of interdependent networks that could represent a determining factor for the production of multiple benefits (ecological-environmental, but also socio-economic) for the settled communities. Each of these networks performs different functions, which could present mutual conflicts, this is why it is desirable to bring the relationship back to dynamic balances characterized by relationships of synergy and compatibility. It is only in this specific interaction, going beyond the "sector" vision, that we grasp the role of the knowledge of the communities of the internal areas and their awareness of being the only subject in the perennial change of the landscape vision of a place.

Bibliography

Barca F (2018) Politica di coesione: tre mosse. IAI Documents, European Parliament, Bruxelles

Costanza R, d'Arge R, de Groot R, Farber S, Grasso M, Hannon B, Limburg K, Naeem S, O'Neill RV, Paruelo J, Raskin RG, Sutton P, van den Belt M (1997) The value of the world's ecosystem services and natural capital. Nature, 387(6630), 253–260. https://doi.org/10.1038/387253a0

Daily GC (ed) (1997) Nature's services: societal dependence on natural ecosystems. Island Press, Washington, DC

Giacomelli M (2023) Ecologies of cohesion: an ecological perspective on territorial cohesion through the lens of landscapes as social-ecological systems. PhD Thesis, Doctoral course in architecture, design, planning—curriculum sustainable urban planning (XXXIV cycle), Scuola di Ateneo di Architettura e design "E. Vittoria", University of Camerino

Giacomelli M, Sargolini M, Felipe-Lucia MR (2024) Including the perspective of stakeholders in landscape planning through the Ecosystem Services co-production framework: an empirical exploration in Le Marche, Italy. Regional Environmental Change 24:24. https://doi.org/10.1007/s10113-024-02184-w

Magnaghi A (2014a) La biorégion urbaine. Petit traité sur le territoire bien commun. Éterotopia France, Paris

Magnaghi A (2014b) La regola e il progetto. Un approccio bioregionalista alla pianificazione territoriale. Firenze University Press, Firenze

MEA – Millennium Ecosystem Assessment (2005) Ecosystems and human Well-being: synthesis. Island Press, Washington, DC

Osti G (2006) Nuovi asceti. Consumatori, imprese e istituzioni di fronte alla crisi ambientale. Il Mulino, Bologna

Santolini R (2008) Paesaggio e sostenibilità: i servizi ecosistemici come nuova chiave di lettura della qualità del sistema d'area vasta. In: Riconquistare il Paesaggio. La Convenzione Europea del Paesaggio e la conservazione della biodiversità in Italia. MIUR—WWF Italia, Roma.

Massimo Sargolini
School of Architecture and Design,
University of Camerino, Ascoli Piceno, Italy

Foreword by María R. Felipe-Lucia

The concept of Nature's Contributions to People (NCP) has reshaped how we understand the interdependence between ecosystems and human well-being. Developed through the work of the Intergovernmental Science-Policy Platform on Biodiversity and Ecosystem Services (IPBES), NCP builds on the legacy of ecosystem services while expanding it to include diverse knowledge systems, including indigenous and local perspectives. The origin of the concept of ecosystem services arises from the field of environmental economics or ecological economics, in an attempt to motivate the conservation of nature by appealing to the economic savings that involve the free supply of a large number of services by the ecosystem (Costanza at al. 1997). Given that previous models of nature conservation, such as the creation of protected spaces isolated from people, or regulations and management strategies at the species level, had failed to reverse biodiversity loss and ecosystem degradation, conservation efforts aimed at highlighting the fundamental role not only of biodiversity, but also of the proper functioning of ecosystems to provide goods and services to humanity. Part of the success of this concept lies in its versatility and possibility of being directly translated into economic terms, essential for political decision-making. However, this practice carries the risk of commodification of nature and the biases derived from it, since not all ecosystem services are equally convertible to economic terms. For instance, many regulating and cultural services would not be taken into account in political decisions. Complementary techniques such as the ecological and social valuation of ecosystem services have revealed the great differences between using one approach or another for the valuation of ecosystem services and therefore the importance of combining economic, ecological, and social valuation approaches for decision-making (Martín-López et al. 2014).

In this sense, the concept of ecosystem services has also contributed to extending the understanding of people as part of the ecosystem, reducing the conceptual separation between nature and humans that persists in numerous disciplines (Mace 2014). This worldview has been fundamentally extended through the *social-ecological systems* conceptual framework (or coupled human-nature systems, Liu et al. 2007), which emphasizes that humans cannot be considered isolated from nature but rather constitute an integral part of a complex and coupled system (Berkes

& Folke 1998; Ostrom 2009). This concept highlights the fundamental role that humans play on ecosystems by influencing their state through both positive and negative actions, as well as the benefits that people obtain from the ecosystem in the form of flows of ecosystem services. Numerous conceptual frameworks have contributed to integrating ecosystem services within the framework of social-ecological systems. In particular, the so-called *cascade model* has widely been used to explain how the flow of ecosystem services arises from the biophysical processes and structures of ecosystems that fulfill certain ecosystem functions, and that, in turn, produce ecosystem services (Haines-Young and Potschin 2010). In turn, it opened the debate on whether ecosystem benefits not perceived or recognized directly by people deserve to be considered as ecosystem services. At a certain point, scholars began to suggest that ecosystem services might not derive exclusively from natural systems (or natural capital in economic terms), but rather that their realization often depends on a degree of human contribution—whether cognitive (service recognition), physical, human, social, or financial capital. This understanding has led to the concept of co-production of ecosystem services (Palomo et al. 2016).

Advances have continued to evolve in recent years, especially following the work of the IPBES, established in 2012 and whose objective is to "provide policy-makers with objective scientific assessments on the state of knowledge about the planet's biodiversity, ecosystems and the benefits they provide to people, as well as the tools and methods to protect and sustainably use these vital natural resources" and can be considered the counterpart of the IPCC on issues of biodiversity. IPBES has not only incorporated these scientific demands, developing an integrative and inclusive conceptual framework, but has taken a fundamental leap by coining the new term, *Nature Contributions to People*, which encompasses and therefore aims to replace the concept of ecosystem services. This new term arises from the need to include other concepts and modes of knowledge, such as local and indigenous knowledge, to provide greater effectiveness and legitimacy to the public policies that emerge from these evaluations. For IPBES, it is essential to recognize different types of paradigms regarding the relationship between people and nature and emphasizes the role of culture in these links. For example, many cultures do not identify with the concept of "receiving services from nature" but rather with "being part of nature" (IPBES 2022). In this way, NCP can be both positive and negative and is classified in a non-exclusive way into material, immaterial, and regulating contributions, since, for example, the collection of wild fruits, despite being a material product, can be carried out with recreational, educational, or medicinal purposes (immaterial). Therefore, and especially for decision-making, IPBES encourages a plural assessment of NCP that incorporates different perspectives on the relationship between humans and nature (IPBES 2022). In line with this concept, IPBES emphasizes the importance of considering the different types of values (instrumental, intrinsic, or relational) that nature has for people, as a fundamental motivation for the conservation of the natural environment (Chan et al. 2016) and to support the path toward the sustainability of our society (Pascual et al. 2023), within a safe and fair space (Rockström et al. 2023).

In this context, the book *Ecologies of Cohesion* provides an excellent opportunity to understand the implications of using a social-ecological systems perspective in landscape planning, particularly to assess territorial policies of inland areas. By investigating different NCP aspects of the inland territory Le Marche region in Italy together with a spatial analysis of inland-coastal interdependencies, this book showcases practical applications to landscape management combining social and ecological valuation tools. This book offers an outstanding guide to enhance the sustainable management of inland areas and ensure territorial cohesion by highlighting the unique value of inland areas in the supply of NCP to the whole region and promote the development of place-based, locally adapted management strategies.

Bibliography

Berkes F, Folke C (1998) Linking social and ecological systems: management practices and social mechanisms for building resilience. Cambridge University Press, Cambridge

Chan KMA, Balvanera P, Benessaiah K et al (2016) Why protect nature? Rethinking values and the environment. Proc Natl Acad Sci USA 113(6):1462–1465. https://doi.org/10.1073/pnas.1525002113

Costanza R, d'Arge R, de Groot R et al (1997) The value of the world's ecosystem services and natural capital. Nature 387:253–260. https://doi.org/10.1038/387253a0

Haines-Young R, Potschin M (2010) The links between biodiversity, ecosystem services and human well-being. In: Raffaelli D, Frid C (eds) Ecosystem ecology: a new synthesis. Cambridge University Press, Cambridge, pp 110–139

IPBES (2022) Methodological assessment regarding the diverse conceptualization of multiple values of nature and its benefits, including biodiversity and ecosystem functions and services. Intergovernmental Science-Policy Platform on Biodiversity and Ecosystem Services, Bonn

Liu J, Dietz T, Carpenter SR et al (2007) Coupled human and natural systems. Ambio 36(8):639–649. https://doi.org/10.1579/0044-7447(2007)36[639:CHANS]2.0.CO;2

Mace GM (2014) Whose conservation? Science 345(6204):1558–1560. https://doi.org/10.1126/science.1254704

Martín-López B, Gómez-Baggethun E, García-Llorente M, Montes C (2014) Trade-offs across value-domains in ecosystem services assessment. Ecol Indic 37:220–228. https://doi.org/10.1016/j.ecolind.2013.03.003

Ostrom E (2009) A general framework for analyzing sustainability of social-ecological systems. Science 325(5939):419–422. https://doi.org/10.1126/science.1172133

Palomo I, Martín-López B, Potschin M, Haines-Young R, Montes C (2016) National Parks, buffer zones and surrounding lands: mapping ecosystem service flows. Ecosyst Serv 17:276–283. https://doi.org/10.1016/j.ecoser.2015.08.002

Pascual U, Adams WM, Chan KMA et al (2023) Diverse values of nature for sustainability. Nature 620:552–560. https://doi.org/10.1038/s41586-023-06365-w
Rockström J, Gupta J, Lenton TM et al (2023) Safe and just earth system boundaries. Nature 619:469–478. https://doi.org/10.1038/s41586-023-06083-8

María R. Felipe-Lucia
Instituto Pirenaico de Ecología - IPE - CSIC, Jaca, Spain

Introduction

The exponential growth of human activities has caused an increasing demand for natural resources, raising concern for the consequences on natural ecosystems. Soil degradation, destruction of forest vegetation, and the decrease in natural habitats are triggering environmental changes that would have catastrophic effects for biodiversity and the life of species on planet earth (Sala et al. 2000). The extraction of biomass, fossil fuels, and minerals from the soil has increased over the last decades by 80%, urban areas have doubled since 1992, and agriculture has significantly shifted toward intensive management all over the world (IPBES 2019). The human impact on the Earth's climate, land, oceans, and biosphere is now so relevant that the birth of a new geological epoch is debated under the name of Anthropocene (Zalasiewicz et al. 2011). Yet, the impact of human action is not equally distributed over space. Despite the relatively small surface they cover, cities have a massive environmental impact well beyond their borders. To fulfill their needs, urban areas are linked to surrounding territories in the extraction and consumption of natural resources through a real "ecosystem appropriation" (Folke et al. 1997).

Beyond the urban fabric, many territories are undergoing an unprecedented demographic and economic decline. Excluded from the urban concentration of capital and investment, rural areas are facing a progressive loss of economic opportunities and essential services, making life in these territories increasingly difficult (ESPON 2021). Nevertheless, urban development hardly exists in the absence of a linkage with rural areas (Gebre and Gebremedhin 2019). For centuries, the agricultural landscapes have provided cities with food, energy, and fresh water. Mountain landscapes covered by woodland support the regulation of water flows, capture air pollutants from the atmosphere, and prevent soil erosion. The awareness of this dependency was further amplified during the COVID-19 sanitary crisis, when the rural open spaces ensured recreational opportunities to the urban population constrained in the cities and forced to measures of social distancing (Beckmann-Wübbelt et al. 2021; Derks et al. 2020). The rural area supplies are often considered free gifts of the environment with little human efforts. Yet, the provision of services and benefits from ecosystems require management practices and care to sustain their current and future use (Felipe-Lucia et al. 2020).

To support an equitable prospect for rural areas, several initiatives are underway in Europe aimed at minimizing spatial disparities among regions and avoiding polarization between cities and surrounding territories. The long-term vision for European rural areas 2040 is currently animating a debate in the direction of stronger, more connected, resilient, and prosperous rural areas (European Commission 2021). This and the other actions directed toward reducing the disparities among regions are based on the concept of territorial cohesion, launched in the Green Paper on Territorial Cohesion (EC, 2008) and included in the Lisbon Treaty (2007) as one of the three main pillars of the EU (European Union) Cohesion Policy.

In this frame, the Italian Strategy for Inland Areas (SNAI) represents a major national application of territorial cohesion in Europe (Lucatelli et al. 2022) and aims at reversing the depopulation trend in the "inland areas" of the country. Defined as distant from the delivery of services as health, education, and transportation, the concept of inland areas aims at overcoming the urban–rural dichotomy, rejecting any dimension of town recognized by theory, and following a polycentric reading of the territory (Lucatelli et al. 2019). Since it was launched in 2013, the SNAI has held the merit of placing marginal areas at the center of the Italian public debate, putting the focus on their possible future through the use and regeneration of their ancient cultural and natural heritage (Barca 2022). At a society level, a growing number of grassroots projects are building on this available heritage experimenting with new models of society (Collettivo PRiNT 2022; Giacomelli and Calcagni 2022; Osti 2006). Despite that, inland areas are still described in terms of underdevelopment and marginality. In the context of regional development, planning and governance face the challenge of integrating the ecological value of inland areas in the territorial assessments in order to guarantee a sustainable landscape development (Albert et al. 2014; Bennett et al. 2015).

Aware that the future of inland areas lies in the interplay between human and nature, the motivation of this book relates to the application of the social-ecological framework to the analysis of regional landscapes, to support the "inland areas" discourse with scientific and objective assessments. Through a multidisciplinary perspective, the integration of an environmental perspective in territorial analysis aims to bridge the gap between cohesion strategies and ecological studies. To date, a growing number of researchers are approaching the complexity of social-ecological interaction to support sustainable landscape planning to cope with current and global challenges (de Vos et al. 2019; Fischer et al. 2015). Yet, our understanding of how—and to which extent—human patterns influence and are affected by landscapes characteristics remains unclear.

Within the motivation described above, this book aims at integrating the social-ecological systems perspective in the territorial analysis, exploring regional landscapes as the result of complex interaction between social and natural components. Links within and between systems are investigated through the concept of Nature's Contributions to People (NCP), which builds on the legacy of Ecosystem Services (ES) while expanding it to include diverse knowledge systems, including indigenous and local perspectives. Defined as "all the benefits society derives from ecosystems," this functional concept allows the visualization of multisectoral relationships between society and nature enabling their integration in planning and regional governance.

This book provides territorial policymakers with a rigorous operational framework, while also advancing theoretical and applied knowledge on how the concept of NCP can be integrated into planning. Building on the tools developed within the ES framework, the study adopts an NCP-based approach to capture a broader and more inclusive set of values and knowledge systems. To achieve this, the work pursues three interrelated objectives. First, it investigates the various fields of application of the ES concept within landscape planning, examining how this framework can support a multidisciplinary vision of the landscape that incorporates both social and ecological dimensions. Second, it develops and tests an approach for mapping landscapes as social-ecological systems through the NCP framework, applied to a Mediterranean regional case study. This includes the identification of spatial interdependencies among local systems and the derivation of recommendations for sustainable landscape planning and territorial cohesion policies. Third, it proposes a framework for analyzing the role of social actors within social-ecological systems by applying the concept of NCP co-production. This involves integrating the social dimension of ES, examining stakeholders' dependencies and benefits, and drawing implications for landscape planning in terms of actor relationships and access to decision-making processes.

These themes are developed across four chapters. The first two provide the theoretical foundations of the work, articulated through two interconnected perspectives: an ecological reading of landscapes as social-ecological systems and the conceptual evolution of territorial cohesion within regional policy frameworks. The third chapter presents the empirical application of these concepts to the case study, Le Marche Region, combining spatial analysis with a qualitative investigation into local actors' perceptions and the co-production of values. The final chapter brings these elements together into an integrated interpretation of landscape systems, contributing to a broader reflection on how to reconcile ecological sustainability with territorial equity. In light of these premises, this book offers both theoretical perspectives and operational insights on how ecological reasoning can support more just and sustainable regional development strategies.

Acknowledgements

This book builds upon the outcomes of research conducted during the XXXIV Cycle of the Doctoral course in *Architecture, Design, Planning: Curriculum Sustainable Urban Planning* at the International School of Advanced Studies of the University of Camerino (UNICAM) and during a Research stay at the German Centre for Integrative Biodiversity Research (iDiv). The work was conceived, designed, and developed under my leadership, with guidance and support from several key individuals and organizations.

My supervisor, Professor Massimo Sargolini, supported the overall development of the research through constructive dialogue and critical feedback. My co-supervisor, Dr. María Felipe-Lucia, provided essential guidance in the definition of the methodological framework and in the design and interpretation of the analyses related to the case study.

I acknowledge Le Marche Regional Authorities and Patrizia Giacomin, which provided data and put me in contact with the competent regional offices, Fulvio Tosi e Gaia Galassi for the feedbacks during the indicator identification, and Fabrizio Cerasoli and Alessandro Zepponi, for the assistance. I thank Alessandro Battoni from the CEA in Macerata, the agronomist Oriana Porfiri and the Regional Soil Observatory in Treia. Finally, I wish to thank the local actors in the Fiastra Valley who participated in the focus group and questionnaire.

I would like to express my appreciation to the PhD reviewers, Professors Francesc Barò and Davide Marino, who offered valuable feedback to improve the work, and endless thanks to Megan Lueneburg for the final proofreading and Greta Papaveri for the graphics rework.

I gratefully acknowledge the German Academic Exchange Service (DAAD) for funding my research stay and I am grateful to my current employer, the Department of Architecture and Urban Studies at Politecnico di Milano, for providing a stimulating environment and the opportunity to refine, expand, and bring this work to completion.

Glossary

Ecosystem services (ES): Benefits society derives from ecosystems, classified as provisioning, i.e., food, water, and timber; regulating, i.e., climate, floods, disease, and waste; cultural, i.e., recreational, experiential, aesthetic, or spiritual benefits; and supporting, i.e., including soil formation, nutrient cycling, or photosynthesis. *ES supply* represents the capacity of ecosystems to provide specific goods and services, while the *ES demand* refers to the amount of service desired by a society. *ES bundles* are a set of positively correlated ES (demand or supply) across space and time (MA, 2005; Villamagna et al., 2013; Raudsepp-Hearne et al., 2010).

Inland area: Territories characterized by a significant distance from the main centers offering essential services identified as health, education, mobility, as well as by a high availability of important environmental resources (e.g., water resources, agricultural systems, and forests) and cultural resources (e.g., archaeological heritage and historical settlements) (Barca et al., 2014).

Landscape: The result of a continuous interaction between nature and humans, which have transformed the territory creating specific regional patterns associated with local historical and cultural backgrounds. The European Landscape Convention (2000) defines it as "an area, as perceived by people, whose character is the result of the action and interaction of natural and/or human factors." The definition includes both the physical object and its interpretation by human communities. *Landscape planning* is understood as a forward-looking and participatory process that guides the transformation, restoration, and creation of landscapes, aiming to balance ecological integrity, cultural values, and social well-being (Sereni, 1961; Gambino, 1996; Council of Europe, 2000; Sargolini, 2013).

Landscape as social-ecological system: Anthropic and natural features in a landscape are coupled to the point that they should be conceived as one social-ecological system. Those systems are complex and adaptive, as they are composed of interdependent and interacting entities: social systems adapt to changes in their environment and, as a result, the environment adapts to their changes (Berkes and Folke, 1992).

Natural capitals: The stock of natural resources that provide benefits to people through ecological functions. Natural capitals include both biotic and abiotic components—such as soils, air, water, and living organisms—which sustain life and contribute to human wellbeing. In this book, natural capitals are also understood in relation to cultural dimensions of landscape, recognizing that the values and meanings attributed to nature are shaped by social practices and collective heritage (Costanza et al., 1997; Díaz et al., 2018).

Nature's Contribution to People (NCP): The concept refers to "all the benefits and detriments that people get from their relationships with the rest of the living world". It represents the evolution of the ES concept embracing inclusive and diverse interpretations of human–nature relations, reflecting the increasing involvement of social sciences and other knowledge systems (e.g., indigenous and local) in global environmental science-policy interfaces. The definition can be particularly useful within the *coproduction* framework to understand how benefits from nature to people are understood differently through different cultural lenses. NCP is a central notion in the work of the Intergovernmental Science-Policy Platform on Biodiversity and Ecosystem Services (IPBES) (Diaz et al., 2018; Locatelli et al., 2024).

Territorial cohesion: Promoting balanced and harmonious territorial development between and within countries, regions, and municipalities, as well as ensuring a future for all people, building on the diversity of places and subsidiarity. It enables more equal opportunities, including access to public services for individuals and enterprises, wherever they are located. Territorial cohesion reduces inequalities between disadvantaged places and those with less prosperous prospects and helps all places to perform as well as possible using their own assets through place-based strategies (European Union, 2007).

Territory: Region or geographical area that includes the social and physical (i.e., morphological, geological, and ecological) dimension of the environment. It is the space where communities establish relationships with the land, in clashing and in sharing, driven by cultural, economic, and social trends. Beyond its geographical definition, the territory can be understood as a system of municipalities that cooperate through functional and relational interdependencies, shaping local development trajectories. *Territorial planning* is conceived as a strategic process that governs the interaction between socio-economic dynamics and the physical environment, aiming to enhance collective well-being (Calafati, 2015; Magnaghi, 2010).

Contents

About the Author

Matteo Giacomelli is a planner and environmental scientist currently employed at the Department of Architecture and Urban Studies of Politecnico di Milano. He holds a degree in Environmental Studies from Aalborg University and a PhD in Sustainable Urban Planning from the University of Camerino, with part of his doctoral research carried out at the German Centre for Integrative Biodiversity Research (iDiv).

His research explores the topics of landscape ecology and regional development, blending theoretical insights with practical applications. He is co-founder and president of Inabita Laboratorio Territoriale, where he coordinates Qui Val di Fiastra, a pilot project for local regeneration in the Marche Region, promoted and funded by the Italian Ministry of Culture.

Chapter 1
Landscape Ecology

1.1 Introduction

Landscapes provide a privileged lens through which to read the dynamics of environmental and social change. As both product and process of human–nature interactions, they embody material patterns, symbolic meanings, and power relations that unfold across space and time. This chapter presents landscape not as a static backdrop, but as an active medium in which ecological processes and cultural values are co-produced, perceived and contested. By adopting a landscape ecology perspective, the following paragraphs link biophysical heterogeneity with social complexity, providing an entry point for the analysis of territorial cohesion.

Landscape Ecology has emerged as a distinctive scientific discipline focused on understanding the spatial patterns and ecological processes that structure and transform landscapes over time. It builds on the recognition that heterogeneity, scale, and connectivity are fundamental properties shaping ecological systems across spatial and temporal gradients (Forman and Godron 1986; Turner 1989). The field integrates ecological theory with spatial analysis, emphasizing how the composition, configuration, and structure of landscape elements influence the flow of energy, matter, and organisms. Key concepts include the idea that landscape mosaics—composed of patches, corridors, and matrixes—mediate ecological processes such as species movement, nutrient cycling, and service provision (Turner et al. 2001).

Building on this foundation, this book situates itself within the disciplinary field of landscape ecology, while also engaging with the broader cultural and political dimensions of landscape. It traces the evolution of the concept from esthetic representation to the understanding of landscapes as social-ecological systems, highlighting how this shift has shaped contemporary approaches to the recognition of landscape values. A key milestone in this trajectory is the European Landscape Convention (2000), which formalized the integration of perception, identity, and

M. Giacomelli, *Ecologies of Cohesion*, SpringerBriefs in Geography,
https://doi.org/10.1007/978-3-032-01159-6_1

agency into landscape assessment and policy-making, thereby acknowledging the role of human experience in shaping and interpreting landscape dynamics.

Following this conceptualization of landscape, the book introduces the framework of Nature's Contributions to People (NCP) as a lens for interpreting human–nature interactions. NCP offers a transdisciplinary and culturally grounded approach to understanding the diverse values and relationships that societies establish with their environments. It complements the scientific rigor of landscape ecology by emphasizing the relational, subjective, and plural dimensions of ecological values.

To translate these insights into operational tools for landscape planning, the concept of Ecosystem Services (ES) is employed as a methodological backbone. ES-based approaches provide structured techniques for mapping, bundle identification, supply–demand modeling, and for analyzing the co-production of services across social-ecological systems. These methods serve to ground the conceptual framework in practice, and are applied in this book to the case of the Marche Region, where they support an integrated assessment of ecological functions and social perceptions within territorial systems.

1.2   The Concept of Landscape: From Aesthetic Conceptions To Social-Ecological Systems

Landscapes are the result of continuous interaction between nature and humans, which has transformed the territory creating distinct regional patterns associated with local historical and cultural contexts. Landscape is the making of a society in a given territory (Sereni 1961) and by definition brings back the ambiguity between "real country" and its representation. The planning practice experiences the dualism between the objectifying attempts of earth sciences and regressions to the estheticizing conception of impressionistic and a-scientific subjectivism (Gambino 1996). The holistic approach necessary for analysis thus suffers from the unresolved tension between the objectivity of ecological reality and the subjectivity linked to the reworkings of the local actors who inhabit and thus modify the landscape.

The term Landscape Ecology was introduced by German bio-geographer Carl Troll in 1939 and originated from the convergence of the spatial approach of the geographer with the functional method of the ecologist (Forman and Godron 1986). Landscape ecology emphasizes the interaction between the ecological patterns of a process, focusing on the causes and consequences of spatial heterogeneity across different scales. The heterogeneity is approached within and between scales, focusing on how it influences the management of natural and human-dominated landscapes (Turner and Gadner 2015). Landscape is thus seen as a mosaic of interacting ecosystems, where the landscape ecologist aims to understand their structures, processes, and meanings for society.

The European Landscape Convention (Council of Europe 2000) attempts to resolve the ambiguity between the objectivity of ecological reality and the subjectivity of perception, proposing a definition that integrates these two components:

"Landscape" means an area, as perceived by people, whose character is the result of the action and interaction of natural and/or human factors

In this way, the convention recognizes the complexity of landscapes, incorporating the central role of people and communities in planning and governance (Sargolini 2013). The convention brings along new approaches to landscape assessment, and a new idea of environmental quality, based on the systemic view that embraces all of the components into which it can be divided. A major challenge of planning is to apply this complexity in the analysis, combining objective and quantifiable assessments with more subjective ones, capturing the complexity of the landscape and going beyond the limits of sectoral planning (Sargolini and Gambino 2016). This need is made even more urgent today, when strong global pressures are appearing in regional landscapes, threatening the balance of ecosystems (IPBES 2019).

The interlinked dynamics of environmental and societal change can be addressed and understood through the concept of social-ecological systems. Fischer et al. (2015) underline the capacity of this framework to support the recognition of the human dependence on ecosystems, improving collaboration across disciplines and between science and society. Despite the growing application in the context of landscape planning, critical open challenges are related to the understanding of social-ecological interactions between regions and the interactions among power relations and access rights, leading to open questions in the field of environmental justice (Felipe-Lucia et al. 2015; Fischer et al. 2015).

This book adopts the lens of landscapes as an ideal frame for reading the interaction between ecological and social systems. Figure 1.1 shows how anthropogenic and natural contribution co-produce landscape, and how its perception by the population can shape the process of planning and management of the landscape itself. This circle of interactions is read in this volume through the concept of NCP, which allows to structure and visualize the benefits people derive from ecosystems (the issue is explored in more detail in the next section). The literature further stresses how the landscape unit is the most suitable for the assessment of impacts and variation in environmental quality and it is furthermore appropriate to land use decisions concerning biodiversity conservation and NCP (Nogué and Sala 2018; Tallis et al. 2015; Díaz et al. 2015). In this sense, together with the main goal of territorial cohesion, the research aims at supporting landscape planning toward the integration of the perception and role of social systems in maintaining habitat quality and enhancing biodiversity.

The dual nature of the landscape as both ecological structure and social construct is reflected in the organization of the empirical core of the volume. Chap. 3 combines two complementary components: a spatial analysis based on biophysical indicators, aimed at capturing the objective dimensions of landscape functioning; and a qualitative investigation of local actors' perceptions and values, designed to explore the subjective, socio-cultural construction of place. This integrated approach allows for a reading of landscapes as social-ecological systems, where the material and symbolic dimensions are co-produced and interdependent.

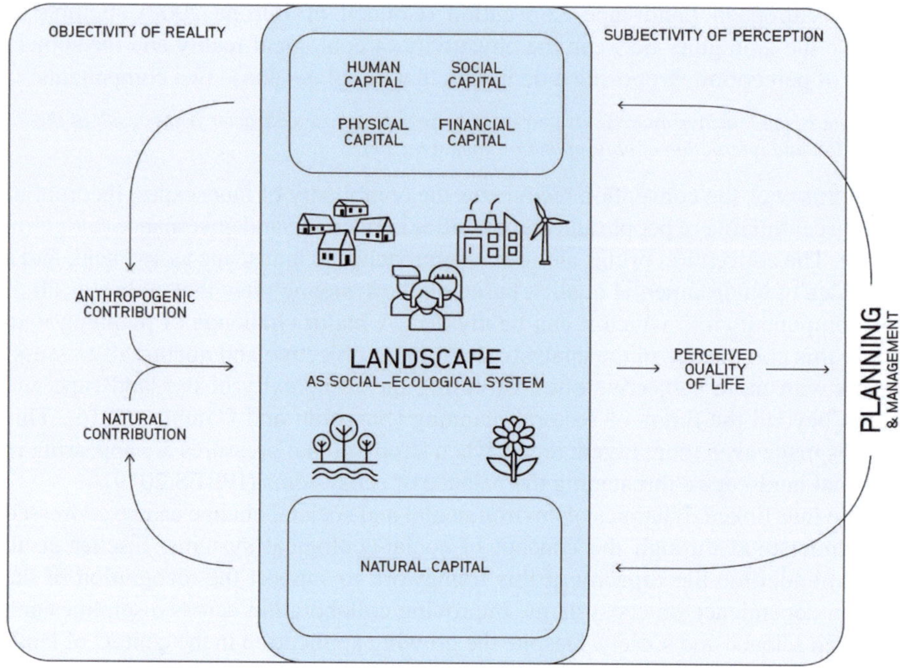

Fig. 1.1 Conceptual framing of landscapes as social-ecological systems adopted in this book

1.3   Framing the Environment through Nature's Contributions to People

The history of our planet can be analyzed in terms of a series of smaller periods of time, referred to as geological eras. Although it has faced periods of significant environmental change, the planet has experienced a period of stability—known to geologists as the Holocene—where humankind lived in harmony with the rest of nature on earth. During the Holocene, the interaction between human and natural activity have coexisted in regional patterns. The regularity of temperatures, the availability of freshwater and biogeochemical flows enabled human civilizations to emerge, develop, and thrive in balance with the environmental assets (Zalasiewicz et al. 2011).

Such stability is now under threat. Since the industrial revolution, human actions have become the main driver of global environmental change. In the so-called Anthropocene, human activities push the Earth outside the environmental stability and planetary boundaries, with catastrophic consequences for the whole globe (Rockström et al. 2023). The capitalist development model requires food, energy, and raw materials for allowing the exponential growth we are experiencing, and this is happening increasingly at the expense of the rest of nature (Haraway 2015). The biosphere, upon which humanity depends, is being altered to an unprecedented degree across all spatial scales. Defined as diversity within species and ecosystems, biodiversity is today declining faster than at any time and human action is proved to threaten more species with global extinction now than ever before (IPBES 2019).

Aware of the human impact on the life on earth, scientists and decision-makers debate over instrumental or intrinsic value approaches to this "environmental issue" (Lele et al. 2018). Within the landscape and territorial planning, the intrinsic approach has been translated and applied through the establishment of protected areas and ecosystem conservation strategies. It has led to the success of policies to protect species habitats in relation to the spatial changes taking place outside them (Chape et al. 2005). Since the creation of Yellowstone National Park in 1872, the extent of protected areas has grown exponentially over the years, and it is still today one of the major environmental measures in terms of biodiversity conservation (see, for example, the EU Biodiversity Strategy target of 30% of land in Europe under legal protection by 2030). Nevertheless, while the state of ecosystems has been maintained in parks and protected areas, beyond the fences we assisted to an unprecedented loss of natural surface area.

Overcoming the perception of nature to be preserved with respect to a generically defined "wild" (Cumming and Allen 2017), this research recognizes the human presence on earth and seeks to analyze the reasons for ongoing injustices and exploitation of nature—in the spatial and social dimension—in order to explore possible ways out. The social-ecological system approach combines the human action on the planet together with the essential role of ecological elements for societal well-being (Binder et al. 2013). In this framework, the Ecosystem Services (ES) lens allows the visualization of how society profits from ecosystems, integrating biophysical, social, and cultural aspects in one assessment. Avoiding the risk of endorsing the reproduction of market logics to environmental goods and services (Gómez-Baggethun et al. 2010), this study escapes from monetary evaluations, while keeping the instrumental power of the ES definition to bridge environmental sciences to an intrinsic anthropocentric discipline such as planning and governance. This utilitarian perspective also challenges conventional wisdoms, including the belief that conservation in planning is based on ethics rather than the benefits society derives from ecosystems (Haines-Young and Potschin-Young 2010).

To embrace the broad range of knowledge systems and stakeholders involved in human–nature interactions, this research adopts the definition of Nature's Contributions to People (NCP) (Díaz et al. 2018). NCP is a transdisciplinary, action-oriented, inclusive, and pluralistic analytical tool for understanding the benefits and detriments people derive from their relationships with nature. It emerged from a global process involving governments, civil society, academic disciplines, and indigenous peoples and local communities. The term "Nature's Contributions to People" is an imperfect translation of concepts debated across multiple languages. NCP focuses on the relationships between people and nature, which can be unidirectional or intricately reciprocal, with nature viewed as having agency (Hill et al. 2021). It supports plural valuations of nature, future scenario evaluations, conservation instrument design, and subnational assessments by civil society (Hill et al. 2021). In this sense, the Platform on Biodiversity and Ecosystem Services (IPBES) has adopted an NCP conceptual framework to organize its reports and communication tasks.

The NCP framework includes and builds upon the literature on ES, recognizing ES as one culturally situated interpretation of people–nature interactions, among others. Unlike the traditional ES approach, NCP embraces a degree of conceptual fuzziness among broad categories, treating it as a feature rather than a limitation. Furthermore, the entities of nature considered as sources of contributions are not restricted to the ecosystem level alone (Hill et al. 2021). While the ES framework confine culture in an isolated category—the Cultural Ecosystem Services—criticized for being difficult to define and operationalize (among others, Chan et al. 2012), in the NCP conceptual framework culture permeates all three NCP groups: regulating, material, non-material and expands the categories for integrating and reflecting social sciences and humanities in assessments of people and nature (Kadykalo et al. 2019).

In the debate following the introduction of the NCP framework, de Groot et al. (2018), responding to Díaz et al. (2018), acknowledged the valuable contributions of many indigenous peoples and local knowledge holders to the ES community. However, Díaz et al. (2018) pointed out that among more than 20,000 ES-related publications indexed in Scopus between 1972 and 2018, fewer than 3% explicitly reference Indigenous and Local Knowledge (ILK), and an even smaller share (~0.2%) approach ES research from the standpoint of indigenous and local worldviews (Kadykalo et al. 2019).

Despite these differences, most ES categories fit well within the generalized 18-NCP classification and NCP methodology builds upon the ES concept in multiple ways (Pascual et al. 2017). Given that this book analyzes social-ecological systems based on local knowledge and community perceptions, it adopts the NCP framework as its conceptual foundation. Nonetheless, the case study applications draw on methodologies developed within the ES literature, which serves as a fundamental reference for the structure of the analysis.

> **Box 1.1 A Social-Ecological Approach: Leveraging Ecosystem Services Tools Within the Nature's Contribution to People Framework**
> This volume analyzes social-ecological systems through the lens of local knowledge and community perceptions, adopting the NCP definition as its theoretical framework. The choice of NCP is deliberate, reflecting its inclusive and holistic approach to understanding the diverse ways in which people benefit from and interact with nature. NCP's recognition of cultural, material, and regulating contributions as interconnected and permeated by cultural dimensions aligns well with the aim to capture the multifaceted nature of human-nature interactions in local contexts.
>
> However, while the theoretical framework is grounded in NCP, the practical methodologies employed in the case studies draw extensively from the

established literature on ES. The ES framework, with its rich body of research and well-developed methodologies, provides a robust foundation for empirical analysis. This includes techniques for mapping ES through indicators, and assessing the impacts of ecosystem changes on human well-being. By leveraging the methodological strengths of ES, the book ensures a rigorous analytical approach. At the same time, the NCP framework allows for a more refined interpretation of results, particularly in recognizing the cultural and subjective aspects of human-nature relationships that traditional ES approaches may overlook.

This dual approach allows the reader to build a comprehensive understanding of social-ecological systems. The case studies thus not only benefit from the empirical rigor of ES methodologies but also from the conceptual richness of NCP, leading to insights that are both quantifiable and qualitatively meaningful. In this way, the book contributes to advancing the field by demonstrating how integrating NCP with ES methodologies can enhance our understanding of human-nature interactions in diverse social-ecological contexts.

In an effort to trace the evolution of ES concept, the Millennium Ecosystem Assessment (MEA) represents the first attempt to systematize and categorize the benefits society derives from ecosystems by grouping them into distinct classes: provisioning (relating to food, water, timber, etc.); regulating (climate, flood control, disease regulation, waste treatment, etc.); and cultural (e.g. recreational, experiential, esthetic, or spiritual benefits). The supporting class, including soil formation, nutrient cycling, or photosynthesis, is classified as basic services and is required to sustain and maintain all the others (MEA 2005). Over the years, the concept has attracted increasing interest and met different territorial disciplines, offering a bridge between science and policy (de Groot et al. 2010a, b). An important operationalization of the concept relies on the cascade model, proposed by Haines-Young and Potschin-Young (2010), which highlights the steps of the flow of contributions from ecosystems to human well-being. The model was further revised through the role of governance in limiting pressures as well as integrating the multiple values ecosystems contribute to human well-being (de Groot et al. 2010a, b; Martín-López et al. 2014). Figure 1.2 presents the framework of the book, discerning the supply side from the demand side. The ES supply represents the capacity of ecosystems to provide specific goods and services, while the demand refers to the amount of service desired by a society (Villamagna et al. 2013). The calculation of budgets between ES supply and demand allows the exploration of mismatches and gaps between different areas (Burkhard et al. 2012). As demand is often not dependent from actual supply within a local system but rather from a larger spatial extent, spatial analyses can help visualizing mismatches among local systems.

In order to identify and characterize local social-ecological systems, the analysis builds on the concept of ES Bundles, defined as set of positively correlated ES

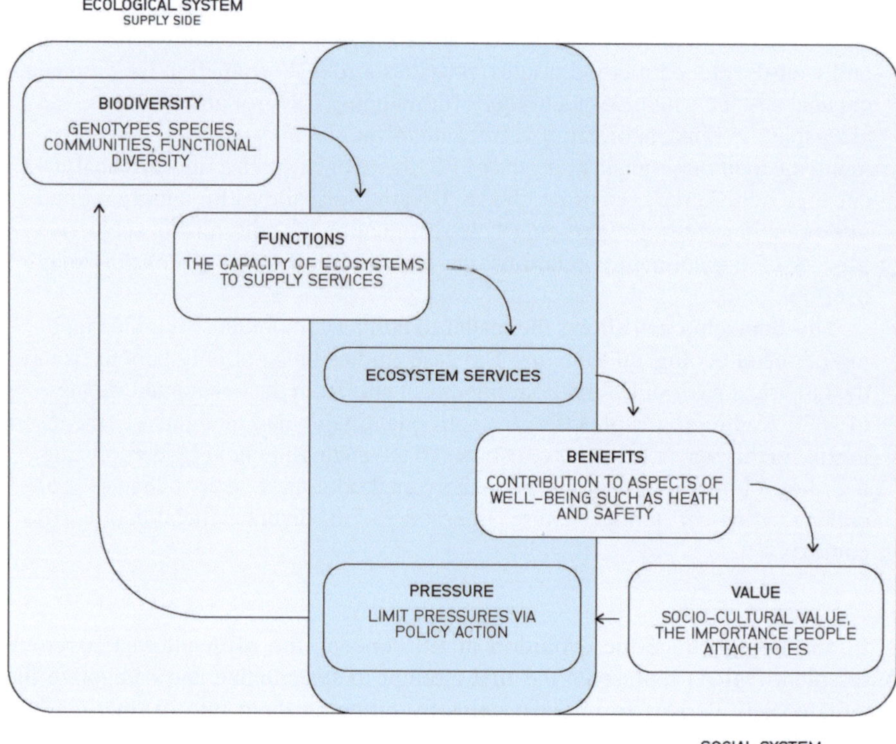

Fig. 1.2 Ecosystem Services cascade model in the social-ecological system. Adapted from Haines-Young and Potschin-Young (2010)

across space and time (Saidi and Spray 2018). This tool is particularly useful for identifying areas of a landscape where ecosystem management has produced exceptional sets of ES and link them to distinct regional socio-economic characteristics (Raudsepp-Hearne et al. 2010). Within Bundles, ES can be positively associated, therefore presenting synergies, or negatively related, representing the case of trade-offs. Trade-offs occur when the provision of one ES is increased at the expenses of another ES. In some cases, this is an explicit choice, but in many others, trade-offs occur without awareness of their taking place. As the role of social actors can shape access to ES and determine which individuals or groups benefit from ecosystems (Felipe-Lucia et al. 2015), the book complements the spatial analysis with a social assessment framed within the concept of ES co-production. This concept stresses how benefits from nature to people do not occur independently but in most of the cases require a significant human contribution. Co-production of ES include several anthropogenic components relating to natural systems, such as motivation or education (human capitals), values and norms (social capitals), machinery and infrastructure (physical capital), and credits or direct payments (financial capitals) (Palomo et al. 2016). Following this rationale, the research includes co-production through the analysis of direct management of service flow (e.g., agricultural activity, forest management) or as users, benefiting and valuing a service (e.g., the preference for a product or a tourist destination, or the ecosystem function toward an environmental risk).

1.4   Ecosystem Services in Landscape Planning: A Literature Review

In order to explore how the Ecosystem Services (ES) framework has been incorporated into landscape planning, the following study presents the results of a literature review of academic contributions published in peer-reviewed journals spanning landscape ecology, spatial and environmental planning, regional development, and sustainability-related disciplines. The objective is to examine how ES have been theorized, operationalized, and applied in planning contexts, with particular attention to their integration into spatial governance systems and their potential to support more sustainable and socially inclusive landscape transformations.

The review focused on three main aspects: the sectors involved in the application of ES, the methodologies adopted, and the extent to which the ES framework has contributed to fostering a multidisciplinary vision of landscape. The review process was guided by clearly defined inclusion and exclusion criteria, and followed a transparent protocol outlining each step of selection and analysis. To complement the qualitative assessment, a co-citation analysis was conducted to identify the evolution of research fields over time and to map the main scientific clusters and conceptual linkages within the ES literature. This combination of systematic review and bibliometric mapping allowed for a comprehensive understanding of how the ES framework has been positioned within academic discourse and practice.

By highlighting both conceptual trajectories and methodological patterns, the review provides the theoretical grounding for the empirical investigation developed in Chap. 3 and informs the methodological design of the case study.

1.4.1   Ecosystem Services in Planning Practice: State of the Art and Current Debates

Over the past two decades, the ES framework has become a central reference for visualizing and analyzing the relationships between ecological functions, social needs, and decision-making processes in landscape planning (Bennett et al. 2015; de Groot et al. 2010a, b). With its capacity to link biodiversity and human well-being, the ES concept is increasingly seen as a useful integrative tool for addressing the complexity of territorial transformation and guiding sustainable landscape planning (Albert et al. 2016; Longato et al. 2021; Ruckelshaus et al. 2015).

The literature highlights how the ES framework facilitates cross-sectoral dialog and supports planning decisions in a range of domains such as land and water management, agricultural policy, conservation strategies, and resilience planning for extreme events (Albert et al. 2014; Sargolini 2013; Sitas et al. 2014). By explicitly naming and structuring the benefits provided by nature, the framework enables a better negotiation of trade-offs among ecological, economic, and social goals (Grêt-Regamey et al. 2017). Yet, despite this potential, applications of ES in planning remain largely confined to isolated interventions—such as evaluating the benefits of afforestation programs (Yu et al. 2018) or implementing payment schemes for water quality improvement (Keeler et al. 2012)—rather than being embedded in integrated territorial strategies (TEEB 2010; Longato et al. 2021).

Recent contributions stress the need for multi-objective approaches capable of reconciling environmental goals with social equity (Benra et al. 2022) and recognize the relevance of social-ecological systems thinking as a means to articulate the co-dependence between society and ecosystems (de Vos et al. 2019; Fischer et al. 2015). Landscapes are increasingly conceived as dynamic, adaptive systems, where ecological and social components are interlinked and co-evolve. At this scale, ES can serve not only as a descriptive tool but also as a transformative framework for redefining planning practices in ways that incorporate power dynamics, governance structures, and social justice concerns (Tallis et al. 2015).

However, the integration of ES into spatial and territorial planning is still limited, partly due to the fragmentation between scientific production and policy implementation (Albert et al. 2014; Bennett et al. 2015). Among the main barriers are the scarcity of tools for operationalizing ES within existing governance systems, the lack of practical knowledge for cross-sectoral integration, and the still partial recognition of the role of culture, perception, and social dynamics in ES assessments (Dick et al. 2018; Geneletti et al. 2020).

Despite these challenges, a growing body of research is expanding the application of ES beyond the boundaries of ecological economics and into more inclusive and participatory planning contexts. A particularly vibrant field of investigation concerns the relationship between cultural values and ES. Recent studies emphasize the importance of subjective dimensions—such as motivation, education, and symbolic meaning—in shaping the ways people perceive, value, and interact with landscapes (Díaz et al. 2018; Palomo et al. 2016). This evolution has contributed to the emergence of more holistic approaches, in which social learning, local knowledge and stakeholder engagement are key components of planning with nature.

Spatial analysis remains a dominant methodology in the ES literature, with a proliferation of tools for mapping ecosystem functions, service flows, and trade-offs (Brunner et al. 2017; Salata et al. 2020). These analyses are often enriched by participatory techniques, including questionnaires, semi-structured interviews, and focus groups, which involve both experts and local communities (Elbakidze et al. 2017; Giedych and Maksymiuk 2017; Kopperoinen et al. 2014). In parallel, the use of multi-criteria decision analysis (MCDA) is gaining traction as a way to support negotiation processes and evaluate conflicting land use scenarios (Langemeyer et al. 2016; Grêt-Regamey et al. 2017).

1.4.2   A Global Research Landscape with Uneven Geographies

As shown in Fig. 1.3, the use of ES in landscape planning has been expanding globally, with a notable increase in the number of publications in recent years. Europe, North America, and China currently represent the most active regions in terms of

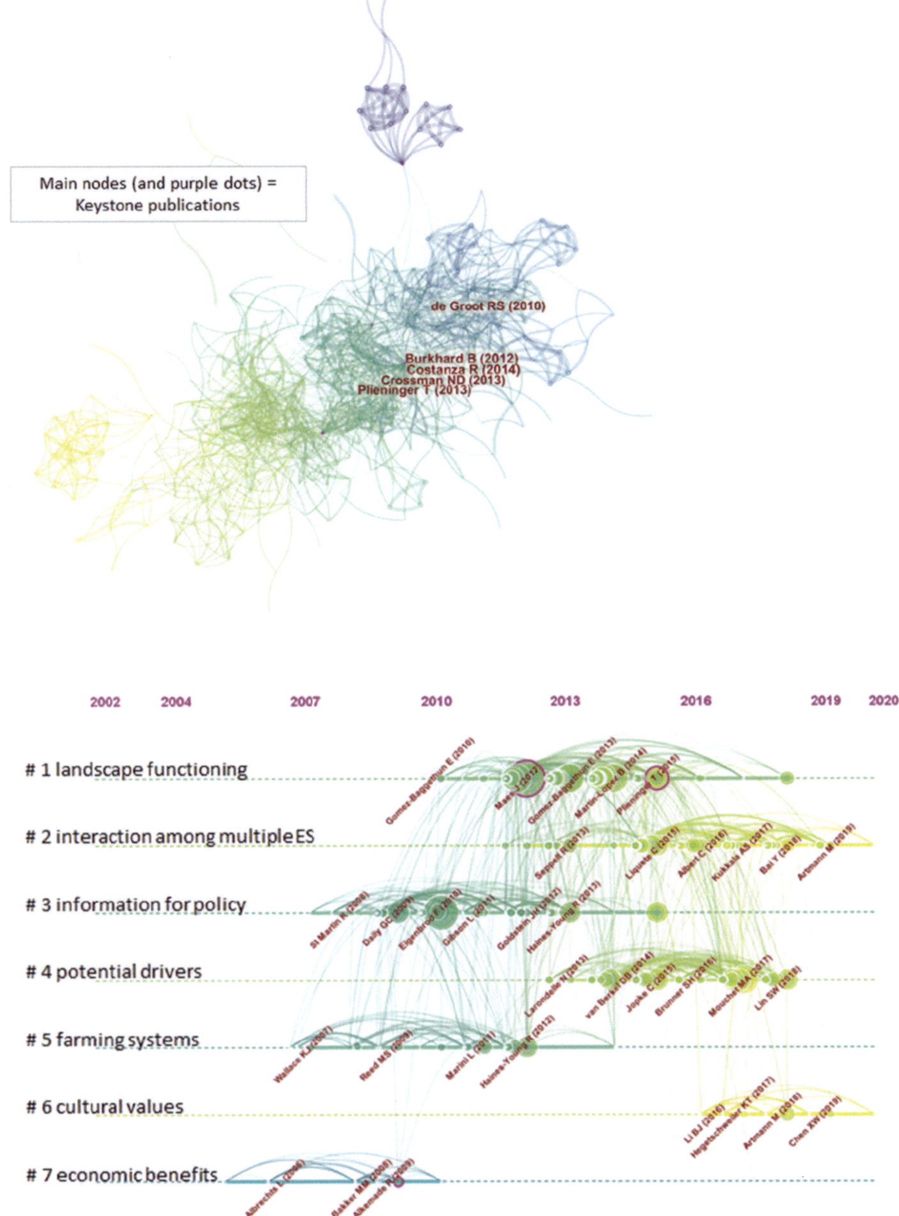

Fig. 1.3 Co-citation networks and clusters of co-citation distributed over time

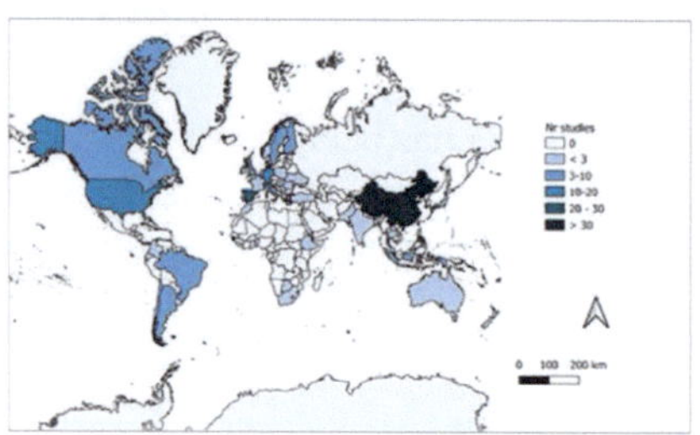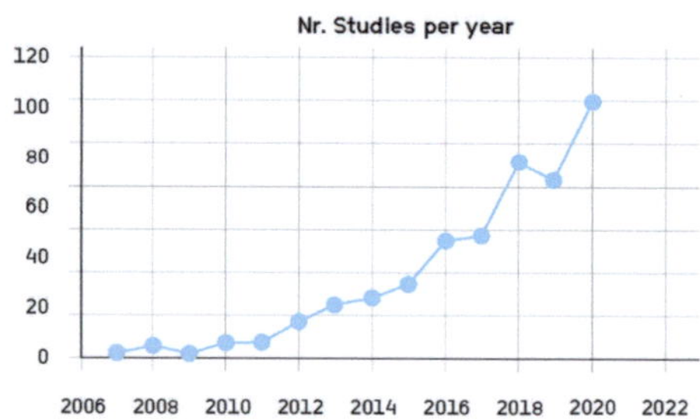

Fig. 1.4 Map of studies localization (case study/study team—first author affiliation) and trend of publications on the topic (number per year)

academic production and policy-oriented application. At the same time, emerging contributions from Latin America, Africa, and South Asia reflect a growing interest in the potential of ES frameworks to support planning in diverse ecological and governance contexts. This expansion is not limited to single case studies: several comparative analyses—both within and across national borders—demonstrate the increasing maturity of the field and the ambition to generalize ES-based planning approaches across different territorial settings (Friedrich et al. 2020; Turkelboom et al. 2018; Momm-Schult et al. 2013).

However, the geographical distribution of research activities reveals important asymmetries. Figure 1.4 shows how a large share of the published literature still originates in the Global North, and many of the studies conducted in Southern regions are led by researchers affiliated with institutions in Europe or North America. This dynamic raises relevant questions regarding knowledge ownership, scientific authority, and the risk of epistemic imbalance in the global development of ES-based planning practices (Spangenberg et al. 2018; Teixeira et al. 2019).

1.4.3 Three Pathways of Integration in Planning Practice

Within the diversity of planning contexts, three main application domains of the ES framework can be identified, corresponding to different degrees of intervention in landscape dynamics (Fig. 1.5). The first, often referred to as **proactive planning**, includes the use of ES in the design of future-oriented strategies and interventions. This includes green infrastructure planning, urban nature integration, and ecological zoning, where ES are used both as design parameters and as criteria for evaluating expected benefits (Baró et al. 2017; Lanzas et al. 2019; Vasiljević et al. 2018). The concept of green infrastructure, in particular, plays a pivotal role in these applications, linking ecological functioning with social needs such as accessibility, recreation, and climate resilience (Giedych and Maksymiuk 2017; Meerow and Newell 2017). Its operationalization in urban and regional planning has been tested in

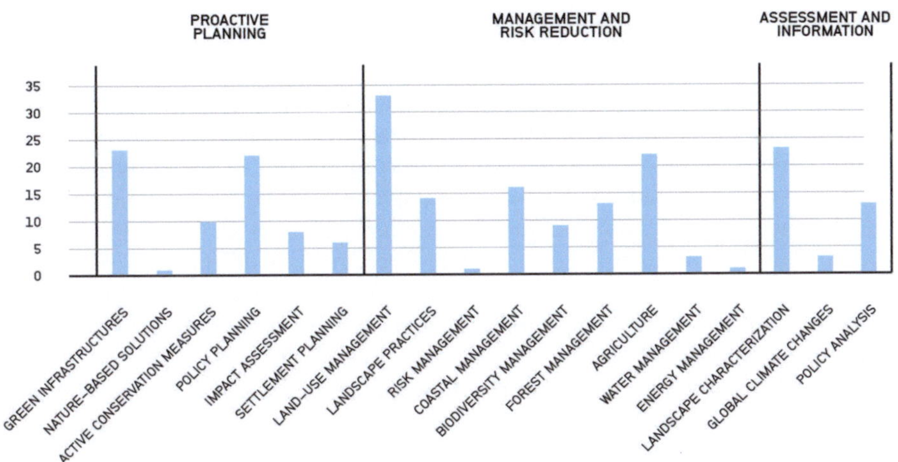

Fig. 1.5 Categories of Ecosystem Services integration in Landscape planning, organized within three macro-categories

multiple contexts, including the city of Dresden (Artmann et al. 2017) and the Lombardy Region (Arcidiacono et al. 2016).

In more strategic contexts, ES-based approaches have supported policy formulation and scenario planning, particularly in relation to the management of urban growth, ecological compensation, and land allocation (Salata et al. 2020; Carmona-Torres et al. 2011; Estrella et al. 2014). Planning strategies have included the use of afforestation targets, compensation schemes for ecosystem conservation, and the application of multi-criteria decision tools to evaluate spatial trade-offs and synergies (Grêt-Regamey et al. 2017).

A second area of application is related to **land use management and risk reduction**, where the ES framework is employed to monitor changes, manage impacts, and guide interventions in evolving landscapes. This includes scenario modeling, trade-off analysis, and participatory evaluations of ES distribution and perception among stakeholders (de Groot 2006; Elbakidze et al. 2018). These tools are particularly relevant in contested or multifunctional landscapes, where different interests coexist and the negotiation of competing claims is necessary.

The third domain focuses on **assessment and information**, encompassing studies that characterize landscape systems in terms of their ecological and social value. Concepts such as landscape livability, multifunctionality, and ecosystem potential are used to develop new forms of spatial representation that go beyond land cover or administrative boundaries (Antognelli and Vizzari 2017; Geneletti 2013; Liu et al. 2019; Zhang et al. 2019). Here, the ES framework supports landscape characterization and zoning, and fosters more inclusive and evidence-based planning cultures.

1.4.4   A Thriving Research Arena

The ES framework has progressively consolidated its role in landscape and spatial planning, developing into a thriving research arena. Recent studies confirm a shift from economic valuation approaches—dominant in the early phases of the field—toward broader frameworks that account for cultural, social and relational dimensions of human–nature interactions (Díaz et al. 2018; Benra et al. 2022). This transition reflects a growing recognition of the role played by anthropogenic components—such as education, motivation, and collective values—in shaping the perception and management of landscapes (Palomo et al. 2016).

Among the methodological approaches, spatial analysis and mapping remain prevalent tools (Brunner et al. 2017; Salata et al. 2020), often combined with non-monetary indicators and participatory techniques including questionnaires, interviews, and focus groups (Giedych and Maksymiuk 2017; Elbakidze et al. 2017; Kopperoinen et al. 2014). Multi-Criteria Decision Analysis (MCDA) also emerges as a key tool to navigate trade-offs in contested landscapes and inform spatial planning decisions (Langemeyer et al. 2016; Grêt-Regamey et al. 2017).

Operationally, the ES framework finds strong applications in green infrastructure planning (Baró et al. 2017; Lanzas et al. 2019), strategic land use scenarios (Salata et al. 2020), and emerging forms of landscape characterization based on livability, multifunctionality, and ecosystem–society relationships (Antognelli and Vizzari 2017; Geneletti 2013; Liu et al. 2019). Particularly promising is the use of ES bundles to identify spatial patterns and co-occurrence of services, thereby supporting the interpretation of landscapes as social-ecological systems (Raudsepp-Hearne et al. 2010; Baró et al. 2017; Queiroz et al. 2015).

Looking forward, promising directions include the integration of intersectional values—such as equity, identity, and cultural specificity—within ES assessment, as well as the development of vernacular indicators that reflect local perceptions and practices (Chen et al. 2019). Future research could explore how these values intersect with ecological processes to support more inclusive and transformative planning paradigms.

In sum, the ES framework offers a fertile ground for both academic inquiry and practical innovation. Its growing methodological diversity and conceptual evolution support the emergence of planning approaches that are adaptive, transdisciplinary, and grounded in the lived experience of places. This work lays the foundation for the analytical path developed in the rest of the volume, informing the selection of indicators, the design of spatial analyses, and the interpretation of landscape transformation through the lens of territorial cohesion.

Box 1.2 The Potential of Ecosystem Services Integration in Planning
Planners and policymakers are increasingly interested in using ES concepts and indicators to support sustainable landscape transformation. The ES framework has shown significant potential to integrate various planning domains, making ES explicit and facilitating discussions on trade-offs between ecological and socio-economic aspects. This ability to bridge different fields of public sector management supports policy decisions in land and water use planning, agriculture policies, conservation strategies, and resilience planning for extreme events.

However, the effective integration of a social-ecological perspective through an ES framework in landscape and territorial planning remains rare. This gap is primarily due to the ineffective interface between science and policy, a lack of practical integration knowledge, and a poor understanding of how to apply ES models in existing planning structures. To bridge this gap, learning from existing practices and ensuring the use of scientific knowledge in decision-making processes are essential.

A literature review can provide a comprehensive analysis of the theoretical and practical integration of the ES concept as a social-ecological tool for landscape planning and management. By identifying application fields of the ES concept in landscape contexts and exploring how it can support a multidisciplinary view of the landscape, the study emphasizes the importance of considering both social and ecological aspects. A systematic literature review offers a rigorous, scientific examination of published material and laying a solid foundation for new insights and analyses.

The review follows a two-phase approach: the first phase involves analyzing co-citation networks to identify research trends, while the second focuses on the selection and quantitative analysis of relevant published papers. This dual approach highlights the main application fields of the ES framework in landscape planning and defines specific categories for its implementation.

Chapter 2
Territorial Cohesion

2.1 Introduction

Over the past century, the world has witnessed a major movement of population toward urban settlements. In the early 1900s, the share of population living in cities was less than 13%, in 1950, this proportion reached almost 30% and in 2009 it surpassed the 50% of the total population (United Nations 2016). This mass exodus from rural territories is expected to continue till 2050 when nearly 70% of the world's population will live in cities. The trend is consistent with shrinking rural regions across Europe, particularly strong in Northern and Mediterranean countries (ESPON 2021). Driven by industrialization and modernization of societies, the concentration of capital and innovation in cities has pushed people to move to urban areas to seek better economic opportunities and thus a higher quality of life.

However, the urbanization process did not happen at no cost. Besides leading to the discussed unprecedented environmental impacts, the rise of urban settlements carried with it a process of marginalization of the rest of the territory, which have suffered strong social and economic impacts (Sørensen 2014). The new urban centralization has led to exclusion from innovation processes and labor markets and areas that shaped cultural and social identity in the past are now politically underrepresented and culturally marginalized (Pittau et al. 2010). Quoting the words of the World Bank (WB 2009), "the concentration of the economic activity is inevitable and usually desiderable for economic growth, but the resulted spatial disparities in welfare are not." Indeed, such trends are widely believed to be a result of political interest and market forces, privileging the highly productive metropolitan areas to the rest of the territory (Medeiros 2016). This sense of abandonment, together with a self-perception as "losers" of the global system, tends to a complex social phenomenon that Stenner (2010) defines as "authoritarian dynamic." Intolerance of diversity, desire for closed communities, demand for strong powers as well as distrust of institutions are some of the social consequences of this tendency.

M. Giacomelli, *Ecologies of Cohesion*, SpringerBriefs in Geography,
https://doi.org/10.1007/978-3-032-01159-6_2

The issue of promoting a more balanced, sustainable territorial development is addressed in Europe through the concept of *Territorial cohesion*, defined by the Lisbon Treaty as the "aim of reducing disparities between the various regions and the backwardness of the least-favored regions" (European Union 2007).

This chapter explores the evolution and current relevance of the concept of territorial cohesion, analyzing the dynamics of spatial inequality across Europe and the political instruments developed to address them. Section 2.2 discusses how territorial disparities have historically emerged and been interpreted within European policy frameworks. Section 2.3 focuses on the Italian context, presenting the National Strategy for Inland Areas (SNAI) as a key case study in place-based policy innovation. Finally, Section 2.4 proposes an integration of ecological thinking into the territorial cohesion debate, advocating for a deeper recognition of the ecological functions and services provided by marginal territories as fundamental components of regional development.

2.2 Territorial Inequalities and Cohesion Policies

Territorial cohesion is a policy and planning concept that emerged in European discourse in the late twentieth century. It refers broadly to the goal of harmonious, balanced development across all regions, so that no people or places are disadvantaged by their location. In 2004, the EU's Third Cohesion Report defined territorial cohesion as ensuring "people should not be in disadvantage by wherever they happen to live or work in the Union." This notion adds a spatial dimension to social and economic cohesion policies, emphasizing spatial equity alongside growth–as Davoudi (2005) explains, it brings the European social model's political tensions "to the fore" and gives them "a spatial dimension." At its core, territorial cohesion seeks to reduce geographic disparities, improve access to services and opportunities in all regions, and foster a more integrated and inclusive European territory.

Integrating and complementing the two pillars of economic and social cohesion (competence of the European Community since the Single European Act, 1986) territorial cohesion applies to disparities between and within countries, regions, and municipalities, and aims to achieving "harmonious development." Derived from France roots,[1] the concept reflected a will to counteract the prevailing tendency of market forces to favor the most competitive and populated regions. Faludi (2007) underlines how this new emerging "EU conceptual novelty" was in support of a European Model of Society, in opposition to the liberal Anglo-Saxon model of development, looking at the equity principle behind territorial cohesion as diametrically opposed to the efficiency principle based on free mobility of labor.

---

[1] It was first discussed by the Assembly of European Regions, under the vice president Robert Savy, and afterwards popularized in the European Commission (EC) by the French commissioner for Regional Policy Michel Barnier, who ensured that territorial cohesion received a mention in the Treaty of Amsterdam, in 1997 (Faludi 2007)

A series of key developments advanced the concept: the European Spatial Development Perspective (ESDP, 1999) introduced polycentric development as a central principle, while the EU's Green Paper on Territorial Cohesion (2008) further reinforced its importance. The Treaty of Lisbon (2007) ultimately recognized territorial cohesion as an official EU objective, placed alongside economic and social cohesion. Since then, it has become one of the three main pillars of EU Cohesion Policy, shaping initiatives such as the Territorial Agenda (2007 and 2011) and subsequent cohesion policy frameworks.

> **Box 2.1 Exploring the Book's title: Ecologies and Cohesion in perspective**
>
> Ecology is the branch of biology that examines the interactions between organisms and their environment, as well as among organisms themselves. Landscape ecology emphasizes these interactions, focusing on the causes and consequences of spatial heterogeneity across different scales. Today, ecology extends beyond traditional studies, incorporating perspectives from the humanities, social sciences, and arts. In sociology, the term refers to the study of relationships between human societies and their environment.
>
> The term "cohesion" literally means the state of cohering or sticking together. In social and organizational contexts, cohesion refers to the bonds or connections that unite individuals into a group, promoting solidarity and unity. The book focuses on territorial cohesion, which aims to unite different areas and regions, supporting less-developed regions through policy approaches.
>
> By combining "ecologies" and "cohesion," the title underscores the inclusion of ecosystem components in efforts to reduce disparities, an aspect often underrepresented in cohesion policies. This definition emphasizes the book's multidisciplinary approach and its openness to different disciplines. Starting from environmental analyses, it aims to explore new meanings of the relationship between human beings and nature.

Though nuanced and sometimes "elusive and ambiguous" (Medeiros 2016), territorial cohesion generally encompasses several core principles. At heart is the idea of spatial justice or equity—all regions (whether urban or rural, central, or peripheral) should have fair access to infrastructure, services, and economic opportunities. Another aspect is balanced development: avoiding excessive concentration of growth in a few core areas (Medeiros 2016). This includes promoting polycentric urban systems (strengthening secondary cities to reduce dominance of a single metropolis). Territorial cohesion also implies better coordination of sectoral policies (transport, environment, economic policy, and so on) to ensure they work together in each place rather than at cross purposes (Faludi 2004). Finally, a territorially cohesive approach values territorial diversity and cooperation—recognizing the unique potential of each region (including disadvantaged or remote areas) and encouraging collaboration across borders and regions. Overall, the concept can be summarized as having three components: reducing geography-related disparities, ensuring coherence between sectoral policies, and strengthening ties between territories. In short, territorial cohesion is a holistic, place-based approach to regional development, aiming for a more balanced and inclusive European territory.

However, the adoption of the Europe 2020 strategy, and with it also the EU Cohesion Policy 2014–2020 follows a "growth" rather than a "development" narrative, including the territorial cohesion within the "inclusive growth" priority, "fostering a high-employment economy delivering social and territorial cohesion" (Medeiros 2016). The undergoing EU political agenda does not place territorial cohesion policy as a main topic of political discussion and the potential for Member States, Regions, and territories to use this tool to increase opportunities for people in remote, mountainous, and the outermost regions is largely unexploited. Barca (2018) stated how cohesion policies are still determined by space-blind decision-making (one-size-fits-all institutional reforms) and public investment driven by corporate decisions. To remedy the inequalities, "compassionate compensations" led to welfarist dynamics toward disadvantaged areas, and this unconditional support is often questioned today as it has not succeeded in addressing structural divergences but often served only to appease anger and potential conflicts (Barca et al. 2012).

2.3 The Italian National Strategy for Inland Areas

Among EU Member States, Italy has pioneered an original application of territorial cohesion principles through the National Strategy for Inner Areas (SNAI). Developed with a place-based approach, it seeks to address within-region disparities through innovative governance and local activation. The attempt to lay the theoretical and operational foundations of this place-based approach to tackle the growing territorial and social inequalities was first embodied in the Report for a Reformed Cohesion Policy[2] (Lucatelli et al. 2022). The document starts from the assessment of how distant cohesion policy is from the ideal model stated in the Lisbon Treaty and propose a set of pillars for developing a new perspective. This proposal aims at a new EU strategic framework for cohesion policy, which includes the implementation and reporting aimed at results, a strengthened governance for the core priorities, the promotion of additional, flexible, and innovative spending as well as experimentalism and the mobilization of local actors (Barca 2009). Four years later[3] these pillars are implemented at the Italian level, within the SNAI. The

---

[2] The report was prepared at the request of Danuta Hübner—EU commissioner for regional policy—by an independent working group, coordinated by Fabrizio Barca, with the aim of drawing up a cohesion policy reform document.

[3] In 2011–2013, during the Monti government in Italy, Fabrizio Barca is Minister for Territorial Cohesion and put in place a new integrated policy called the National Strategy for Inner Areas (SNAI). This policy applies to every region and macro-area in Italy and is directed at recognizing the social and physical fragilities of remote places.

Strategy applies the concept of territorial cohesion to the local scale (within-regions) and aims to reverse the depopulation trend in the "inland areas" classified on the basis of distance indexes from the supply of main citizenship services.

Launched in 2013 under the guidance of Minister Fabrizio Barca, the SNAI represents the most significant national experiment in implementing the European principle of territorial cohesion. It applies a place-based approach to address within-region disparities by supporting development in areas that are significantly distant from essential services such as health, education, and public transport. These "inland areas" (*aree interne*) cover more than 60% of the national territory and include around one-fourth of Italy's population, yet they face processes of demographic decline, service marginalization, and underutilization of territorial capital (Barca, 2014).

Rather than reinforcing the urban–rural dichotomy, SNAI embraces a polycentric reading of space, defining inland areas based on their functional distance from service hubs—municipalities that ensure access to a basic bundle of secondary education, healthcare (first-level emergency hospitals), and railway connectivity. This categorization moves beyond administrative borders, recognizing varying degrees of spatial peripherality (Lucatelli et al. 2022) (Fig. 2.1).

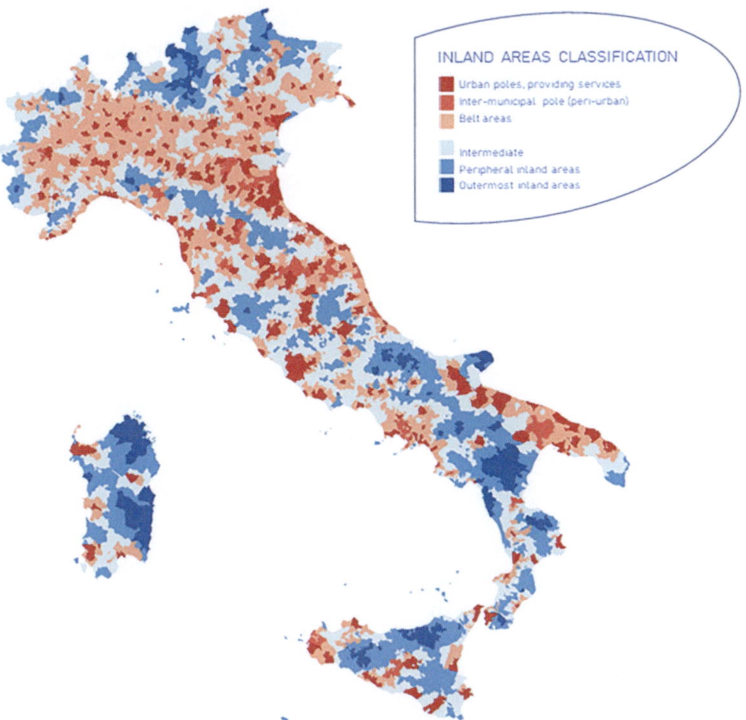

Fig. 2.1 The SNAI inland area municipalities classification. Through the identification of municipal poles delivering services, the SNAI categorize municipalities in relation to travel distance: Urban poles, Inter-municipal poles (peri-urban), Belt areas, Intermediate, Peripheral inland areas, Outermost inland areas. Source: ISTAT, 2018

The strategy introduced a dual intervention model: (a) guaranteeing adequate access to essential services, and (b) co-designing local development projects with municipalities and local communities. These projects include interventions for the active protection of the territory, valorization of cultural and natural resources, renewable energy, sustainable agriculture, and traditional craft systems. From the outset, SNAI emphasized demographic reversal as its ultimate goal, seeing population stabilization and rejuvenation as key indicators of territorial resilience and future viability.

The operationalization of SNAI reflects the broader shift toward negotiated governance models, where national ministries, regional authorities, and local actors collaborate through Program Agreements to define place-based strategies. In addition to introducing institutional innovations, SNAI fostered inter-municipal cooperation and community participation, often for the first time in traditionally marginalized areas (Barca 2022). Mechanisms such as the Federation of Projects have further enhanced collective learning and strengthened a sense of agency within local territories.

Despite these advances, several challenges emerged during implementation. Bureaucratic inertia, delays in fund allocation, and particularly the insufficient integration of environmental dimensions have constrained the strategy's transformative potential. The exclusion of the Ministry for the Environment from the national steering group limited opportunities to embed ES and biodiversity considerations into local development plans (Pierantoni and Sargolini 2021). Moreover, while SNAI stimulated inter-institutional dialog, the emergence of the National Recovery and Resilience Plan (PNRR) has in some cases redirected political attention toward short-term expenditure, overshadowing the strategy's long-term vision for territorial regeneration (Pazzagli, 2023).

Nonetheless, by re-centering the value of territorial diversity and social capital, SNAI has helped to reframe peripheral areas from spaces of abandonment to vital contributors to environmental stewardship. Building on this foundation, future revisions could strengthen the socio-ecological dimension of territorial cohesion by systematically integrating an NCP approach, thus recognizing rural and mountainous areas as key providers of climate resilience, ecological functions, and landscape identity.

2.4 Integrating an Ecological Perspective in Territorial Cohesion

Recent contributions to the literature on territorial cohesion have emphasized the need to move beyond a one-size-fits-all model, advocating instead for flexible, context-sensitive frameworks for regional development. Two main theoretical approaches have gained particular prominence within the European policy landscape. The first is the Territorial Approach to Regional Development (TARD), which promotes place-based policies adapted to the socio-economic and ecological specificities of each region (Noya and Garmendia 2012; Martínez-Fernández et al. 2015). The second is the Smart Specialization Strategy (S3), focusing on leveraging local assets and knowledge systems to foster innovation and economic diversification (Foray et al. 2009; McCann and Ortega-Argilés 2015). Both approaches contribute to a more integrated vision of territorial cohesion, highlighting the necessity of accounting for the interplay between economic, social, and environmental dimensions.

However, despite these advances, critics have noted that ecological processes are often treated as secondary considerations rather than integral elements of regional development strategies. This oversight is particularly problematic in areas that provide essential ecological benefits to society. In response to this gap, the book proposes a complementary socio-ecological perspective on territorial cohesion—one in which environmental dynamics are positioned as fundamental to territorial equity and resilience (Chamusca 2023). This approach reframes so-called "marginal" areas as ecological keystones, vital for regional balance and long-term sustainability.

A growing body of literature has emphasized the importance of integrating environmental and sustainability concerns within the broader framework of territorial cohesion. Among the four dimensions identified by Medeiros (2016), particular attention is given here to the Environmental and Sustainability dimension (Fig. 2.2),

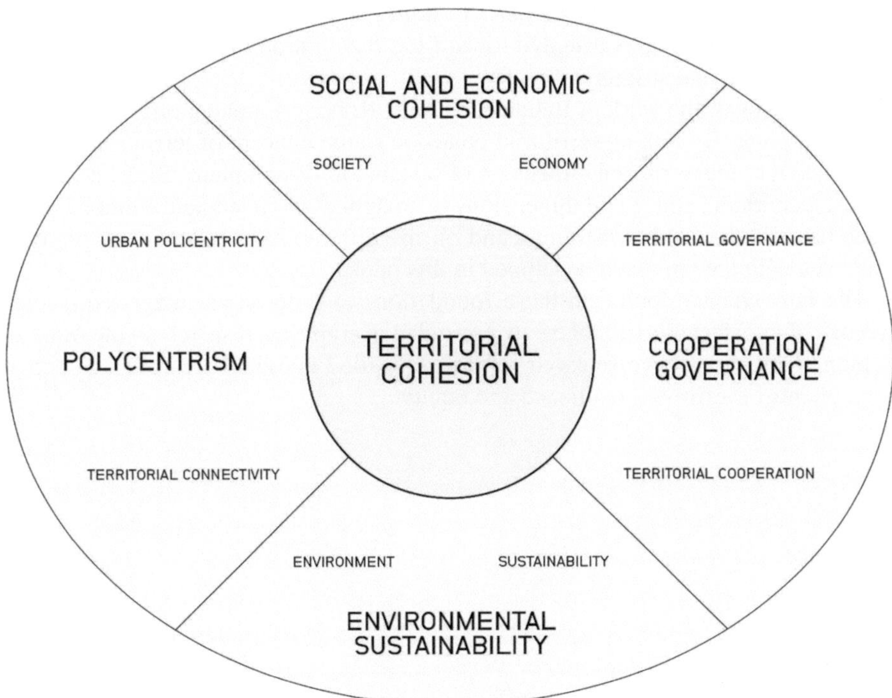

Fig. 2.2 The four dimensions of territorial cohesion. Source: Medeiros (2016)

which has gained increasing relevance as ecological pressures and climate objectives have progressively shaped the European policy agenda. With the support of the ESPON program (among others TEQUILA, INTERCO), the Territorial Agenda 2030 (2020) defines two overarching objectives, a "Just Europe" and a "Green Europe." In the latter, the report specifies the goal to protect common livelihoods and shapes societal transition, especially supporting the development of nature-based solutions as well as green and blue infrastructure networks linking ecosystems and protected areas in spatial planning, land management and other policies. Reference is made to COVID-19 pandemic, that has changed policy making and future development outlooks. As implications and policy responses vary across territories due to different conditions, the pandemic shows that "territories matter" and are highly interdependent (Territorial Agenda 2030 2020).

In the pursuit of an ecological perspective on cohesion policies, this book highlights the role of inland systems in providing essential benefits to society through the lens of NCP. Previous studies have shown that the growth of urban settlements has historically relied on the exploitation of natural capital, much of which originates from inland and rural areas (Gebre and Gebremedhin 2019). Yet, this dependency is largely overlooked within the neoclassical economic paradigm, which tends to treat environmental benefits as positive externalities or "free gifts of nature" (Gómez-Baggethun et al. 2010). By integrating biophysical and socio-cultural assessments, the research seeks to make visible the spatial injustices embedded in dominant urban development models. In doing so, it proposes new elements for spatial cohesion strategies oriented toward the recognition of biodiversity and the human practices that sustain it.

In this context, the work of Italian economist Roberto Camagni offers a valuable reference point. By linking territorial cohesion to the concept of territorial capital, he framed it as the territorial dimension of sustainability, complementing its technological, economic, and social dimensions (Camagni 2008). Camagni's emphasis on place-based assets and constraints, and on the differentiated potential of regions, resonates with the approach developed in this book.

The next chapters build on these foundations to propose an integrated, socio-ecological interpretation of cohesion, grounded in empirical research on the Marche Region. This perspective seeks to reframe so-called marginal territories as active components of territorial resilience and equity.

Chapter 3
Ecologies of Cohesion in Practice: Le Marche Case Study

3.1 Introducing the Practice

This chapter presents the empirical core of the book, structured around two complementary studies conducted in Le Marche Region. Through spatial analysis tools and social investigation methods, the studies operationalize the concepts introduced in Chaps. 1 and 2. investigating the territorial articulation of Nature's Contributions to People (NCP) framework, operationalized through tools and methodologies derived from the Ecosystem Services (ES) literature. The first adopts a regional-scale spatial perspective, while the second focuses on a local scale social investigation of stakeholders' perceptions and roles.

The first study develops a spatial analysis based on mapping indicators to assess the distribution of 12 representative ES, selected in collaboration with regional authorities. These services are used as proxies to operationalize the NCP approach through established ES mapping techniques. The analysis was conducted at the municipal scale, given the availability of data and the relevance of this institutional level for political decision-making. Indicators of supply and demand were mapped and used for two complementary analyses: (1) a cluster analysis to identify ES bundles, and (2) a budgeting exercise to assess spatial interdependencies among municipalities. These outputs were then integrated with socio-economic data to delineate landscape units that reflect the functional organization of social-ecological systems. At the same time, ES budgets were interpreted using the urban–inland classification from the Italian Strategy for Inner Areas (SNAI), allowing for the visualization of regional gradients and interdependencies. Statistical analyses supported the exploration of trade-offs, synergies, and associations among variables, generating insights for sustainable landscape management and informing territorial cohesion strategies.

The second study complements the spatial analysis with a qualitative, actor-centered investigation in the Fiastra Valley, a local system of Municipalities within the region. Here, the NCP framework is applied through participatory tools—focus

M. Giacomelli, *Ecologies of Cohesion*, SpringerBriefs in Geography, https://doi.org/10.1007/978-3-032-01159-6_3

groups and semi-structured interviews—with local stakeholders. The analysis explores how different actors contribute to the co-production of ES, examining the types of capital involved, perceived benefits, dependencies, and access to decision-making processes. Attention is also given to stakeholder self-perception and collaborative dynamics across governance scales. This perspective allows for the integration of local knowledge, values, and power relations into the analysis of social-ecological systems, and supports a more inclusive vision of landscape planning.

3.1.1 The Case Study: Le Marche Region and the Fiastra Valley

The selection of the case studies was linked to the availability of data and the opportunities for engagement with authorities and local stakeholders. The choice of the regional scope is related to *VAUTERECO,*[1] a research project in support of the definition of Le Marche Regional Sustainable Development Strategy. The collaboration with the project allowed for easier exchange of information and collection of non-public regional data. The choice of the local case study of Fiastra Valley is linked to the activities of *Inabita Laboratorio Territoriale,*[2] that is promoting a valley-scale regeneration process together with the six local municipalities facing the valley. The selection of this case study not only allowed for easier involvement of local stakeholders but also gave the study a chance of concrete implementation in local planning. The two case studies are shown in Fig. 3.1.

Le Marche (9.344 km²) is a central-Italy Region bounded on the east by the Adriatic Sea and on the west by the Apennine chain. Characterized by a high landscape diversity, it is described as a good example case of the Mediterranean regions (Bevilacqua 2013). The region comprises a western mountain area, a central hilly belt characterized by rural landscapes and mostly small settlements, and a coastal area consisting in an urban continuum along the Adriatic seaside. From the coastal strip, urbanizations expand along the valleys, where faster road connections penetrate the inner part of the region toward major towns and allow links with the western side of the Apennines. In Le Marche, the Regional Environmental Landscape Plan (*Piano Paesistico Ambientale Regionale*—PPAR) is configured as a territorial plan, and refers to the entire regional territory, including natural areas with cultural

---

[1] *VAUTERECO* (Assessment of Urban and Spatial Assets for Community Resilience) is a research project funded by the Marche Region, aimed at supporting the development of strategies to enhance territorial resilience, particularly in areas affected by the 2016–2017 Central Italy earthquakes. The project involved universities, research institutions, and local authorities in the collection and analysis of environmental, socio-economic, and spatial data to inform sustainable and inclusive regional planning.

[2] Inabita Laboratorio Territoriale is a research and project-based center based in the Val di Fiastra that coordinates. Qui Val di Fiastra, a pilot regeneration project of the Municipalities of Ripe San Ginesio, Loro Piceno and Colmurano, funded by the Italian Ministry of Culture. The initiative combines participatory research, cultural programming, and strategic planning to support local administrations in long-term territorial development. More information at: www.quivaldifiastra.com

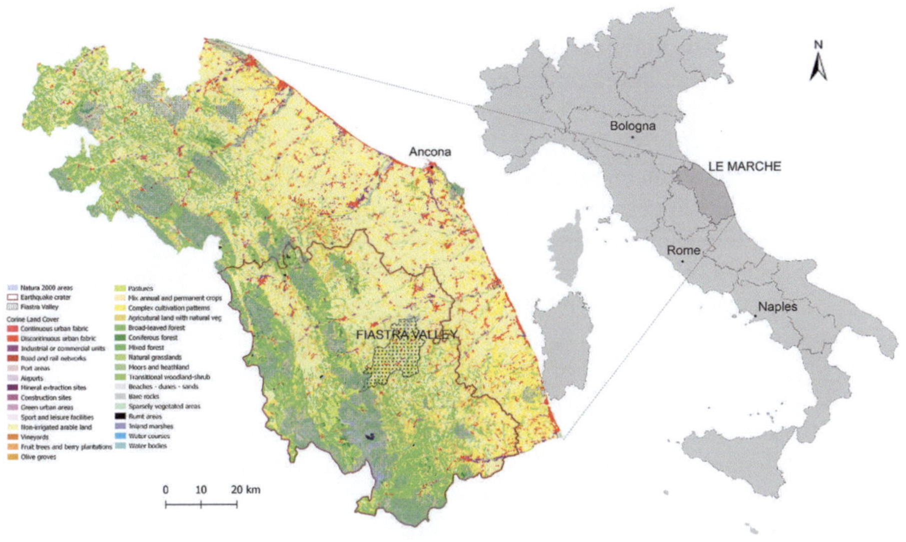

Fig. 3.1 Le Marche regional case study and the Fiastra Valley. Beside the information on land use, the map includes the protected areas of the Natura 2000 network and the limits of the Central-Italy 2016–2017 earthquake crater, that is, the area affected by the earthquake, with special reference to damage incurred by people and properties

value, but also urban or degraded areas. NCP and ES information can provide relevant insights for the integration of a Green Infrastructure perspective into future landscape planning tools. In terms of regional cohesion, Le Marche inland areas, as other Mediterranean areas of Europe, have been for years at the center of the international academic and political debate, concerning development strategies against the phenomena of depopulation and economic decline. The phenomena already described in the Theoretical framework are combined here with the effects of the severe earthquake that hit Central-Italy in 2016–2017. Causing 41.000 displaced persons, 388 injured and 303 dead, the event had catastrophic effects on the built heritage but also in exacerbating the dynamics of abandonment. While the physical reconstruction has barely started, regional governance debates today over the best approaches to support life in the area.

Box 3.1 Lessons from Le Marche Polycentric Territory
Le Marche case study offers valuable insights for understanding urban gradients through a "Polycentric territory" characterized by a spatial arrangement where multiple urban centers, or poles, are integrated with rural and peripheral areas, forming a network of interconnected settlements. In Le Marche case study urbanization expands from the coast into the valleys, facilitated by road connections linking inland areas to major centers.

Inland areas in Italy, including Le Marche, have been central to international debates and development policies, such as the National Strategy for Inland Areas (SNAI), which addresses issues of depopulation and economic decline. The 2016–2017 Central Italy earthquakes exacerbated these problems, causing significant displacement and damage. This situation makes the Region a significant case study for territorial cohesion, addressing the marginalization and inequalities resulting from urbanization.

The social-ecological systems approach is applied to Le Marche through the NCP framework local systems described through 24 ES indicators as well as 9 socio-economic indexes. Peripherality of local systems are ranked according to the SNAI classification, based on travel distance from main service poles.

Le Marche Region offers a meaningful example of how polycentric territorial structures can inform cohesion-oriented planning. This configuration contrasts with monocentric or highly polarized models and provides a useful lens for analyzing interdependencies across spatial gradients. This method can be applied to other Mediterranean and European regions characterized by polycentricity and can support areas often overlooked by territorial policies that prioritize socio-economic assets over ecological functions.

In the inner Marche Region, the Fiastra Valley (43° 9′ N, 13° 50′ E) is a sparsely populated rural district, covering a hilly area characterized by ancient settlements and agricultural land crossed by the Fiastra river. It is located at the foothills of the Apennine mountains and included in the Earthquake crater, that is, the area affected by the earthquake. The land cover consists mainly of arable land, with few forests covering mostly riparian and high-inclination areas. Its territory is constituted by six municipalities and counts about 11.764 inhabitants for 181.2 km² land.

Social actors involved in the study were selected according to five stakeholder groups: *Production* includes workers from the agriculture sector, agronomy, and local producers; *School and research* includes school teachers, university students, ecology experts, and a representative from environmental centers; *Culture and commerce* includes tourist managers, hotel, agritourism, and restaurant owners, café and a local shop owner; *Planning and administration* includes members of the municipality, engineers, architects, and planners, and the local water distribution company; *Society* includes members of local associations, artists, family doctors, local recreationists, and other inhabitants.

The Fiastra Valley, as well as other areas in the region, can be read with the category of "local system," conceived by Calafati and Mazzoni (2009). This interpretation addresses and interprets local systems as new poles present in the mental maps of individuals, visible in the spatial patterns of transactions, and in the spatial organization of settlements. The authors further underline how this polycentric territory is today without government and without a strategy, characterized by socio-economic imbalances as well as evolutionary potential (Calafati and Mazzoni 2009). By adopting this scheme, this book attempts to recognize the design of a new territorial organization, proposing a frame of inter-municipal connection and regulation.

3.2 Social-Ecological Systems and Regional Interdependencies

The process of urbanization together with the ongoing exponential growth of human activities is leading to increasing marginalization and inequalities between regions (Rockström et al. 2023). While cities are mostly associated with economic success and power (Sassen 2018), the inland areas—defined as remote from the delivering of services such as health, education, and mobility—are undergoing a process of economic decline and depopulation which increasingly subordinate them to urban centers (ESPON 2021). In 2009 the urban population surpassed the number of people living in rural areas and the United Nations (2016) forecasts a global increase to 60% by 2030.

Inland areas are crucial in terms of goods and services they provide to human society, such as food and clean water, or climate and hydraulic regulation (Gebre and Gebremedhin 2019). The great role they serve for the well-being of society was especially evident during the COVID-19 sanitary crisis, when the urban population recognized the value of inland areas for the availability of open spaces and recreational opportunities (Beckmann-Wübbelt et al. 2021; Derks et al. 2020). These benefits are increasingly assessed under the definition of ES, useful to analyze urban–rural interdependencies (Gebre and Gebremedhin 2019) and to integrate the link between natural and social systems in landscape planning (de Groot et al. 2010a, b). Within this framework, ES supply represents the capacity of ecosystems to provide specific goods and services, while the demand refers to the amount of service desired by a society (Villamagna et al. 2013). The calculation of budgets between ES supply and demand allows the exploration of mismatches and gaps between different areas (Burkhard et al. 2012). As demand is often not dependent from actual supply within a local system but rather from a larger spatial extent, spatial analyses can help visualizing mismatches among local systems.

Several studies assessed regional spatial interdependencies from an urban perspective, evaluating the effects of city growth on the rest of the region (Peng et al. 2020) or analyzing the ES supply–demand along the urban–rural gradient (Baró et al. 2017). However, the dichotomy between urban and rural fails to integrate the role of inland areas, which host a great share of socio-cultural heritage (Antrop 2005). Inland areas in fact consist of a multiplicity of small urban settlements that have shaped societies over centuries of human–nature interplay (Blondel 2006). This is particularly evident in Mediterranean region, one of the world's biodiversity hotspots (Myers et al. 2000), where social-ecological balances are today threatened by increasing urban pressures and depopulation of local systems leading to two opposite scenarios: agricultural intensification in peri-urban and accessible areas as well as abandonment of peripheric and mountainous areas (García-Llorente et al. 2012). While intensification of agricultural production may impact on the provision of regulating and cultural services (Felipe-Lucia et al. 2014), natural revegetation following the abandonment can help improve some ecological functions and services, such as erosion control and water quality (Bruno et al. 2021). However, abandonment of traditional agricultural and forest management practices (often associated with low-intensity and semi-subsistence) also bring important consequences in the loss of local traditional knowledge and sense of place (Iniesta-Arandia et al. 2015).

Such interaction and trade-offs are often tackled in the literature through the frame of social-ecological systems, which can help analyzing how social systems adapt to changes in their environment and, in turn, how the environment adapts to social changes (Binder et al. 2013). In the context of landscape planning, it is crucial to understand what kinds of social-ecological systems are present in a landscape, as different configurations of societal interactions with nature are characterized by different resource use patterns, human well-being outcomes, development trajectories, and potentials for environmental traps or collapse (Cumming et al. 2014). To approach this complexity, the concept of ES bundles can be effective in identifying areas of a landscape where ecosystem management has produced exceptional sets of ES and can be linked to distinct regional social-ecological characteristics (Raudsepp-Hearne et al. 2010). Many studies have applied this concept in order to characterize landscapes according to ES features (Baró et al. 2017; Peng et al. 2020; Queiroz et al. 2015; Quintas-Soriano et al. 2019). However, the field still lacks evidence on how additional socio-economic factors affect the resilience and sustainability of ES bundles and which social-ecological characteristics are related to the supply and demand of ES (Bennett et al. 2015; Rieb et al. 2017).

This study explores landscapes as social-ecological systems combining spatial features with patterns of ES supply and demand. Building upon the methodology of Raudsepp-Hearne et al. (2010) in the construction of bundles, it further analyzes the role of socio-economic indicators on ES patterns. Parallelly, through budgeting, spatial interdependencies are investigated in the support of policy-making for inland areas (Barca et al. 2012; Gebre and Gebremedhin 2019). Figure 3.2 provides a visual illustration of the two aims of the study: a) to develop and test an approach to

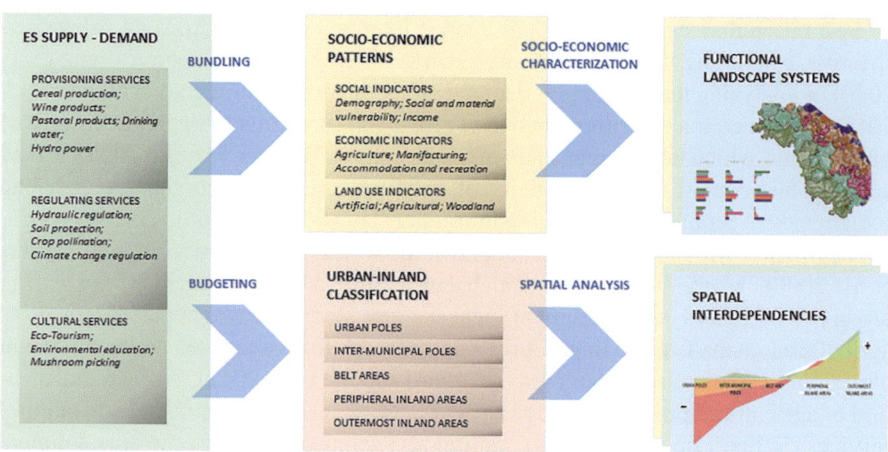

Fig. 3.2 Structure of the study flow. Areas represent layers of information while arrows the flow of analysis

map social-ecological systems in a Mediterranean case study and b) to highlight spatial interdependencies in order to support balanced and just development between urban poles and inland areas. In this sense the study intends to provide a tool for the sustainable landscape development on the one hand and for regional cohesion strategies on the other.

3.2.1 Patterns of ES Supply and Demand

Methods for Ecosystem Services Mapping

The identification of context-relevant ES indicators is the first step for the analysis of social-ecological systems (Burkhard et al. 2012). In this regional analysis, we selected 12 ES and 9 socio-economic features across 227 municipalities of Le Marche Region. The services were classified according to the CICES classification as five provisioning services, four regulating services, and three cultural services (Haines-Young and Potschin-Young 2018). Table 3.1 provides the list of ES and relative indicators of supply (i.e., ecosystems' capacity to deliver ES) and demand (i.e., the amount of ES required or desired by society) (Villamagna et al. 2013).

Table 3.1 List of selected Ecosystem Services and indicators for supply and demand. When not specified, the unit is the same for Supply (S) and demand (D). Abbreviations: *ppl* people, *Inhab* inhabitants, *Lic* licenses

Ecosystem service	Unit	Supply indicator	Data source	Demand indicator	Data source
P1 cereal production	Tons/year/km^2	Cereal production per municipality	ISTAT 2019	Cereal consumption per municipality	ISMEA 2020
P2 wine products	Tons/year/km^2	Wine production per municipality	ISTAT 2019	Wine consumption per municipality	OIV 2014
P3 pastoral products	L/year/km^2	Cheese production per municipality	ISTAT 2019	Cheese consumption per municipality	CLAL 2020
P4 drinking water	1000 m^3/year/km^2	Water catchments by aqueducts per municipality	"Piano Regolatore Acquedotti" Marche	Water delivered by municipal networks	ISTAT 2019
P5 hydro power	Gwh/year/km^2	Hydroelectric nominal production of local power plants	SIGERI 2020	Housing and industrial electricity consumption	TERNA 2019

(continued)

Table 3.1 (continued)

Ecosystem service	Unit	Supply indicator	Data source	Demand indicator	Data source
R1 hydraulic regulation	K (0–100) (S) Km^2/Km^2 (D)	Water retained on total rainfall (1-CN)	SCS curve number method (CLC 2018)	Area at hydraulic risk on total municipal area (%)	ISTAT 2017
R2 soil protection	Tons/ Km^2/year	Potential sediment retained by soil	InVEST sediment delivery ratio model (CLC 2018)	Annual soil loss by water erosion	EU dataset (JRC 2016)
R3 crop pollination	K (0–1) (S) K (D)	Relative pollination potential of municipal surface	EU dataset (MAES, 2010)	Crop dependency by pollinator	Capri model (ESTIMAP 2013)
R4 climate change regulation	Mg CO^2/ km^2/year	CO2 absorption per municipality	Emissions data (Marche 2019)	CO2 emissions per municipality	Emissions data (Marche 2019)
C1 eco-tourism	Km/km^2 (S) ppl/ km" (D)	OSM footpaths mapped per municipality	OSM (2021)	Nr. Hosts at eco-tourist facilities	Marche dataset (2019)
C2 environmental education	K/km^2 (S) Inhab/ Km^2 (D)	Nr. Education centers per municipality	Marche dataset (2019)	Population in schooling age per municipality	ISTAT 2019
C3 mushroom picking	Tons/km^2/ year (S) Lic/km^2 (D)	Suitable surfaces per municipalities	Corine Land Cover (2018)	Nr. Licenses per municipality	Marche dataset

Each municipality was assessed using the same set of ES, considering average values per spatial unit. To allow comparisons across municipalities, ES indicators were spatially standardized by dividing each value by municipal area. Before analyzing the data, measurements obtained for each indicator were further normalized by maximum and minimum values, excluding outliers by substituting values differing from the mean by more than 2x standard deviation.

Spatial distribution maps for each indicator were produced using QGIS 3.10.11 A Coruña to visualize patterns of supply and demand. An analysis of hotspots and coldspots was conducted, highlighting municipalities with the highest and lowest values of ES supply and demand. These values were calculated by summing the normalized scores of the 12 ES.

The hotspot–coldspot analysis was complemented by an assessment of ecosystem multifunctionality, evaluated through Simpson's Diversity Index using the R software package vegan (Oksanen et al. 2017). This allowed us to measure the diversity of ES provision within each municipality.

Despite the high variability in municipal area sizes (ranging from 3.85 to 272.08 km²), the use of municipalities as spatial units allowed for high data availability and an emphasis on the role of local administrations in spatial transformation processes (Barca 2009; Calafati 2015; Felipe-Lucia et al. 2014).

Provisioning Services: Within the Provisioning Services category, supply refers to tangible goods produced and demand refers to the actual consumption by the population. For Regulating Services, supply refers to potential ecosystem functions, while demand is related to the risks arising from service shortages under current environmental policies. Regarding Cultural Services, both supply and demand reflect potential values, such as the existence of recreational infrastructure or the presence of service seekers in the region.

P1 Cereal production indicators are related to actual production (supply) and consumption (demand). The supply value at municipal scale is derived by the spatialization of total regional production on the hectares of municipal surfaces dedicated to agriculture (agriculture census). The demand is mapped by spatializing the total consumption by population density. The assumption here is that the consumption is equally shared among the regional population. The *P2 Wine production* was mapped similarly to P1, with the difference that the demand (consumption) is spatialized according to population over 16 years old, the age below which wine consumption is prohibited in Italy. The *P3 Pastoral products* refer to the production of cheese from goat and sheep farming while the demand considers the available data of total cheese consumption. *P4 Drinking water* relates to the total water delivered by Le Marche regional aqueducts. Data on supply consists of the total regional water extracted, spatialized by the punctual catchments weighted by the maximum uptake allowed by the Regional Aqueduct Master Plan (Regione Marche, 2014). Data on demand relies on the actual water delivered by the regional network, thus excluding losses (ISTAT, 2019). *P5 Hydro power* indicators are related to the electricity produced by regional water plants (supply) and the actual consumptions for residential and industrial purposes (demand). The supply accounts for the nominal power of plants, which are added up within the municipalities they belong to (SIGERI, 2020), while demand spatializes average regional consumptions by the number of inhabitants (domestic consumption) and companies active of each municipality (industrial consumption).

Regulating Services: *R1 Hydraulic regulation* refers to the role played by the natural ecosystem in retaining water during rain events and therefore decreasing the regional flood risk. The supply is mapped according to the curve number (CN) parameter, which predicts direct runoff or infiltration from excess rainfall (USDA, 1969). Input information for the CN method are the hydrologic soil group database

(Regione Marche) and Corine Land Cover (2018). Indicators for demand are the areas of municipalities in flood risk mapped by the Regional Hydrogeological Plan (PAI) as P3 and P4. The ES *R2 Soil protection* relates to the capacity of ecosystems to prevent soil loss and therefore nutrient loss. The supply follows the indicator of Potential sediment retained, mapped through the model InVEST Nutrient Delivery Ratio (Sharp et al. 2014). As for the demand, the ES was mapped according to the data available at EU scale through the RUSLE 2015 model, resolution 100 m (Panagos et al. 2015).

R3 Crop pollination refers to the capacity of ecosystems to provide habitats for pollinator species, essential for ecosystem functioning as well as for several agricultural activities. The supply map uses available data at EU scale available for the year 2010, assuming minor changes in the potential of land cover cells to provide crop pollination. The indicator follows a relative scale between 0 and 1 and is based on input information including relative suitability of land cover cells to host pollinator populations, the availability to provide floral resources and the average activity of bees as a result of climatic variation (Zulian et al. 2013). The demand is instead related to the dependency of cultures by pollination, for which the cultures of each municipality were linked to dependency values (Joint Research Centre et al. 2014). The mapping of the ES *R4 Climate change regulation* assumes the net zero CO2 emission as the ideal condition, in line with the EU's commitment to global climate action under the Paris Agreement (EU, 2021). In this sense, the supply indicator assesses the total CO2 absorption by ecosystems and the demand accounts the emissions of the production sector. The data follow a report of Assessment and quantification of atmospheric emissions in Le Marche Region (UNIVPM 2019).

Cultural Services: Cultural services embrace non-material benefits people obtain from nature and include Environmental education as well as Eco-Tourism and Mushroom picking, together considered within the definition of Recreational services.

C1 Eco-Tourism refers to the capacity of ecosystems to provide opportunities for tourism directed toward natural cultural environments, intended to do sport activities, and observe wildlife. Supply maps follow the indicator of existing trekking tracks, mapped in the Open Street Maps platform (2021). The kilometers of tracks are summed for each municipality and spatialized within the municipal areas. On the other hand, the demand is related to the presence of tourist structures, considering camping sites and Agritourisms for coastal municipalities (in order to exclude seaside tourism) and all tourist structures for the rest of the municipalities. *C3 Mushroom picking* demand indicators follows the distribution of licenses per municipality, provided by Le Marche regional authorities. On the other hand, the supply values were calculated following the methodology by Marino et al. 2014, assuming an average annual production of 1.5–3 kg mushrooms per hectare of forested surface, though Corine Land Cover Classes 231, 243, 244, (weight 1) 311,

312, 313, 321, 322, 324, (weight 2) under 2000 m altitude and with slopes lower than 80%. We considered the DEM Marche (100 m) and extract elevation, then intersect with CLC forested areas.

Finally, *C2 Environmental education* was considered as the process of learning from nature both in formal academic programs, to complement traditional forms of learning, and in less-than-formal settings, such as through the interpretive services offered at natural parks or farms. The indicator of demand is related to population at the age of schooling (6–16-year-old) living in each municipality. The supply map is related to the location of Environmental Education Centers (CEA) and Didactic Farms (DF), recognized by Le Marche Region. Data per municipality was provided by Le Marche regional authorities and weights were assigned as follows: three points per each CEA and one per each DF.

Results: Spatial Patterns and Territorial Gradients

This section presents an overall summary of the ES patterns of supply and demand (Fig. 3.3), while the detailed results per each single ES can be found in the supplement material.

Within the provisioning services, the supply of *P1 Cereal production* highlights a main agricultural strip in the mid-low hilly area from south to north. No values are recorded in the south-western mountain areas while lower values concern the rest of the inland areas. Differently, *P2 Wine products*, shows a spotted concentration of production in areas recognized as DOC and DOP, such as: Rosso Conero (in the area of Ancona), Rosso Piceno (in the low Ascoli Piceno Province) and Verdicchio (area of Jesi and Matelica). Looking at the ES related to *Pastoral activities*, P3 records productions mostly in the mountain part in the southwest and in the north of the region, while low or no values are mapped along the coast. Moving on to water resources, *P4 Drinking water* highlights different profiles for the northern and southern part of the region: the north offers a rather uniformed distribution with hotspots corresponding to the main water withdrawal points, while the south shows a strong supply from the mountain areas. Similarly, *P5 Hydro power* highlights hotspots of production in the mountain area in the southwest of the region and other municipalities along the rivers but no major differences are recorded within the regional area. In terms of ES demand, all the provisioning services show the highest values in the coast and in the main urban poles, where both population density and consumption are higher.

Regulatory service maps are linked by a consistent and marked supply in the upland and forested belt. As for the *R1 Hydraulic regulation* and *R2 Soil protection*, the maps show high values for the mountain areas while low values are mapped in low hilly fields. The *R3* map refers to the relative pollination potential and follows the presence of forests and trees in the inner belt of the region. The same applies for

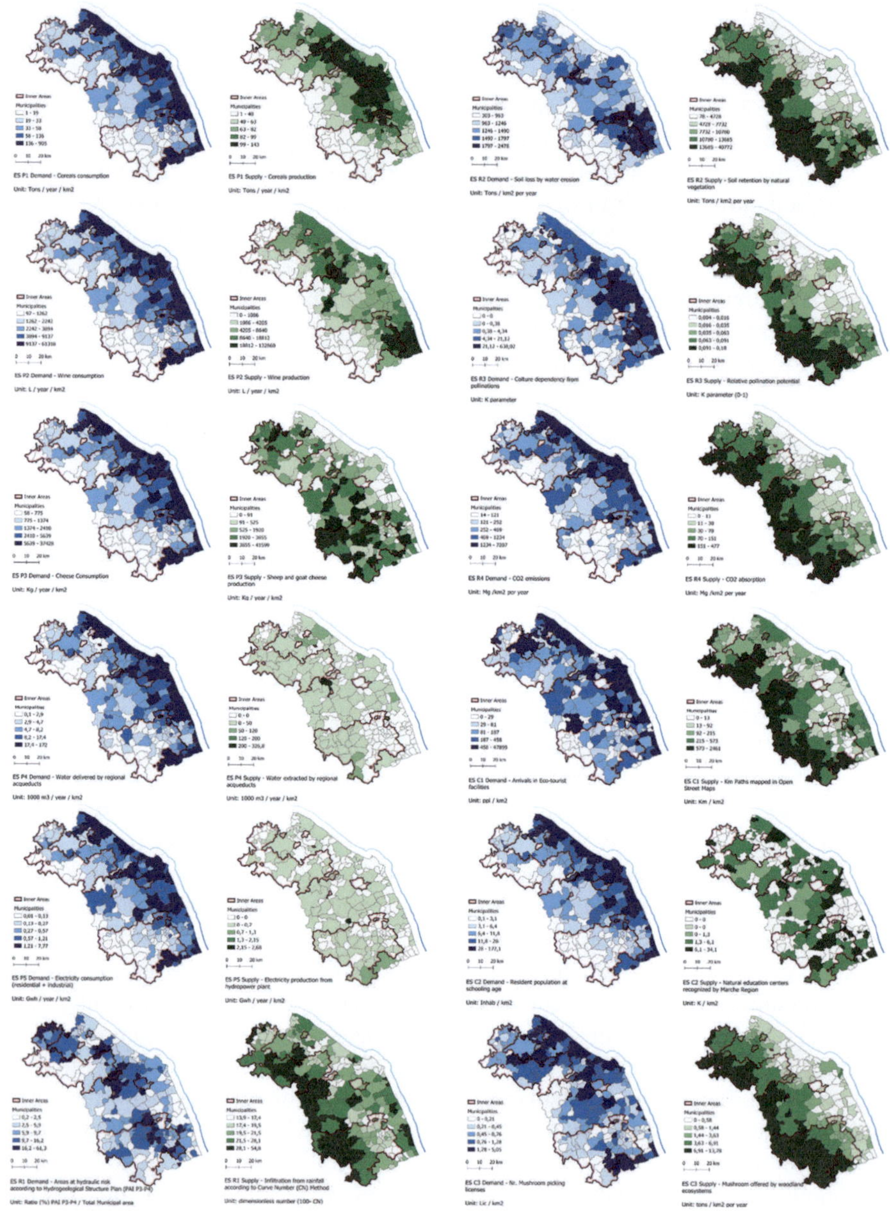

Fig. 3.3 Maps of Ecosystem Services supply (green) and demand (blue). Single higher-definition maps can be found in the Appendix A

the absorption of CO_2. As for the demand, *R1* and *R2* highlight spots of higher pressure, in hilly areas characterized by land instability. *R3* reveals the dependency of cultures from pollination especially in the hilly area devoted to fruit trees and

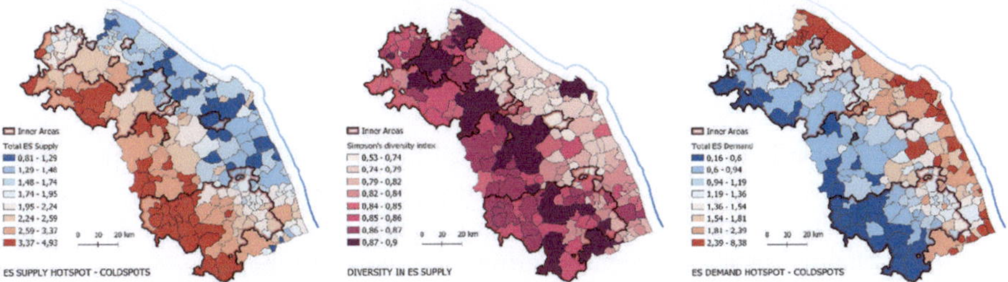

Fig. 3.4 Hotspots–coldspots of Ecosystem Services supply (left), diversity in Ecosystem Services supply (center), hotspots–coldspots of Ecosystem Services Demand (right)

partially oilseeds. *R4* relates to the emission of CO_2, with the map showing higher values for the municipalities of the coast, as well as the first hill belt, hosting most of the regional manufacturing activities. Interesting to note is the concentration of *R4* demand in the footwear production sector located in the central south part of the Region.

Regarding the cultural services, *C1 Eco-tourism* relates to the indicator of available hiking paths and shows higher values in mountain municipalities and protected areas. *C2 Environmental education* reveals an equal spatial distribution of supply connected to the Regional Education Centers and didactic farms. Finally, the supply of *C3 Mushroom picking* is related to geographical condition allowing habitats for mushrooms and has higher values in forested mountain areas coinciding with the inner belt. As for the demand, *C1* and *C2* refer mainly to the coastal areas, while *C3* gives higher values in various territorial hotspots in the region.

The hotspots–coldspots analysis explores the aggregated values of ES supply and demand, highlighting the areas which mostly provide services and the ones who mostly demand them. Figure 3.4 suggests a strong inequality in terms of demand–supply following the inland-coastal gradient. Hotspots of supply can be found in the south-western municipalities, partly coinciding with the Sibillini National Park, and the northern part following the Apennine chain. The municipalities with low supply values are found in the coastal areas and in the first hill range, with the latter presenting the lowest values. Simultaneously, the coastal municipalities present a very high demand, with a declining gradient toward the mountains.

Together with the hotspots–coldspots analysis, we mapped the multifunctionality of each municipal area by calculating the Simpson's diversity in the ES supplied. As shown in Fig. 3.4, the main multifunctional landscapes do not always coincide with the hotspots of ES provision, but primarily cover the upper hills and piedmont areas. Other municipalities characterized by high multifunctionality are those hosting protected areas along the coast, where provisioning services are combined with cultural and regulating services. On the other hand, low multifunctionality is associated with the first hill range, coinciding with the area of strong cereal production.

3.2.2 Territorial Bundles

Methods for Ecosystem Services Bundling

The bundling of ES allows grouping municipalities according to similar profiles of ES supply and demand, highlighting spatial patterns of co-occurrence. This step provides a functional basis to interpret regional landscape systems and their socio-ecological characteristics. To identify different types of ES bundles, we conducted a cluster analysis using the Factoextra package in R (Kassambara and Mundt 2020). We classified municipalities into clusters based on similar combinations of both ES supply and demand values, using K-means clustering algorithm which minimizes the variability withing the groups. A principal component analysis (PCA) was carried out to identify the main explanatory factors and distribution of the 12 ES across the municipalities. We used Rose wind chart to facilitate the visualization of ES demand and supply characteristics. The R code can be found in Appendix C.

Results: Reading the Landscape through Functional Clusters

The ES bundle analysis allowed to group the 228 municipalities of Le Marche Region in five clusters of ES supply–demand characterized by their supply–demand patterns (see Table 3.2), which matched the coastal-inland gradient.

Bundle 1 ("Urban coast") includes 22 municipalities corresponding to the main urban settlements of the coast, together with the small municipalities in the highly urbanized "Tronto Valley," in the southern part of the Region. The Bundle is characterized by high ES demand, which reaches the greatest mean value for all the ES except R1 Hydraulic regulation and R3 Climate change regulation (the only ones

Table 3.2 Bundles mean values for indicators of ES supply and demand

Bundle	Demand											
	P1	P2	P3	P4	P5	R1	R2	R3	R4	C1	C2	C3
B1 urban coastal	0,92	0,92	0,92	0,85	0,83	0,14	0,34	0,13	0,86	0,33	0,95	0,75
B2 cropland	0,33	0,33	0,33	0,27	0,38	0,25	0,45	0,14	0,31	0,11	0,36	0,38
B3 cropland at hydraulic risk	0,13	0,13	0,13	0,12	0,14	0,47	0,78	0,22	0,13	0,02	0,13	0,16
B4 mosaic cropland forest	0,09	0,09	0,09	0,08	0,10	0,43	0,53	0,02	0,09	0,02	0,09	0,27
B5 mountain forests	0,03	0,03	0,03	0,05	0,03	0,16	0,21	0,00	0,05	0,01	0,03	0,14
Bundle	Supply											
	P1	P2	P3	P4	P5	R1	R2	R3	R4	C1	C2	C3
B1 urban coastal	0,48	0,32	0,11	0,22	0,12	0,37	0,17	0,20	0,11	0,21	0,32	0,07
B2 cropland	0,69	0,17	0,07	0,10	0,06	0,26	0,19	0,12	0,06	0,10	0,17	0,07
B3 cropland at hydraulic risk	0,61	0,47	0,25	0,03	0,09	0,21	0,40	0,32	0,13	0,08	0,26	0,16
B4 mosaic cropland forest	0,46	0,07	0,27	0,06	0,05	0,31	0,45	0,53	0,45	0,25	0,14	0,49
B5 mountain forests	0,14	0,02	0,23	0,23	0,10	0,73	0,78	0,80	0,77	0,72	0,16	0,89

lower than 0,30). The supply values are generally low, with the exception of P1 Cereal production reaching a mean value of 0,48. Interestingly, the provision of C2 Environmental education, mapped through the indicator of education structures officially recognized by the Region, presents the highest value among the five clusters (0,32).

Bundle 2 includes 58 municipalities and is named "Cropland" due to the highest values in P1 Agricultural products (0,69). It consists of hilly rural units together with sub-urban municipalities also located along the coast, but with lower population density than the ones of Bundle 1. In terms of ES Demand, it displays moderately high values for all the ES except for P4 Drinking water, R1 Hydraulic regulation, R3 Climate change regulation, and C1 Eco-tourism (lower than 30). In terms of ES Supply, it presents a similar condition as for the urban coastal, with a higher value in P1 (0,69) and slightly lower values for the other ES.

Bundle 3 ("Cropland at hydraulic risk") includes a similar number of municipality than Bundle 2 (57) and differs from it mainly for the presence of a strong demand for R2 Soil protection and, to a lesser extent, for R1 Hydraulic regulation. Except for the three regulation services, all the other ES Demand present lower values than the first two clusters (all below 0,20), mostly related to the lower population density. In terms of ES supply, the slight decrease in P1 Agricultural products is accompanied by a sharp increase in P2 Wine products. This suggests a possible connection between the cultivation of Wine trees and hydraulic instability and, interestingly, the greater demand for hydraulic protection is also associated with greater supply of the service.

Bundle 4, named "Mosaic cropland forest," groups those municipalities ($n = 51$) located in the high hills and in the piedmont areas of the regions. In terms of ES demand, values drop for almost all ES except C3 Mushroom picking (0,27) and the regulatory services. Those services decrease compared to Bundle 2, but remain high with a value of 0,45 for R1 Hydraulic regulation and 0,53 for R2 Soil protection. With respect to the supply, the decrease in P1 agricultural products [which anyway presents a moderate level (0,46)] is combined with an increase in all regulatory services (all higher than 0,30), together with C1 Eco-tourism (0,25) and C3 Mushroom picking (0,49).

Finally, ***Bundle 5*** is called "Mountain forests" ($n = 40$) as it clusters municipalities characterized by high altitude and a great amount of woodland and natural areas. This area hosts only a few large urban settlements and agriculture is absent or minor. This bundle shows by far the highest supply values for all regulating services (all above 0,70) and recreational services as C3 Mushroom picking (0,89) and C1 Eco-tourism (0,72). Services related to water reveal the highest supply of P4 Drinking water (0,23) and the second highest for P5 Hydropower (0,10). In terms of ES demand, the values are the lowest for all the services.

The PCA analysis illustrates the bundles composition in a two-dimensional graph, with Dim1 and Dim2 explaining 39% and 19,1% of the total variance respectively. Explaining most of the differences, Dim1 relates to the altitude gradient, from high mountain areas to low coastal areas through the hilly municipalities in the central part of the region. The gradient is also linked with the decreasing population density from mountain to coast and the consequent demand for ES. Associated to

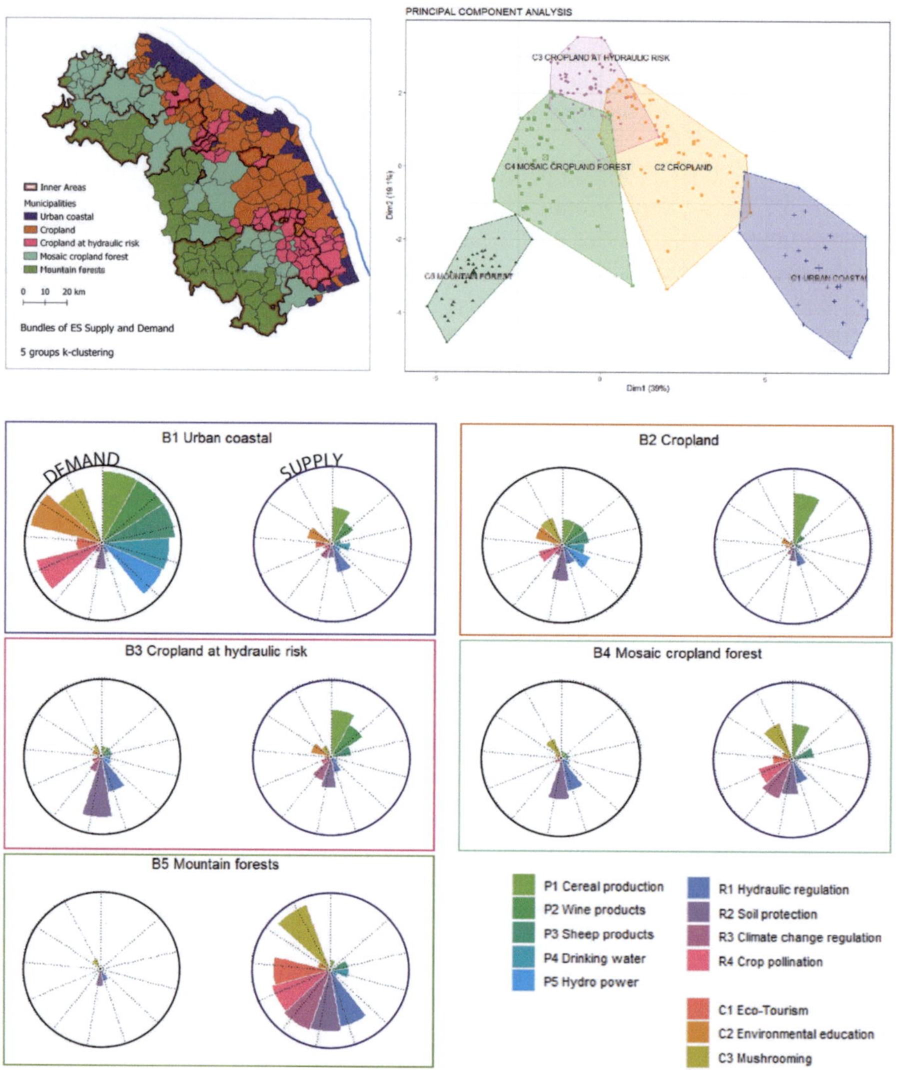

Fig. 3.5 Regional map of Ecosystem Services Bundles (top-left), Principal component analysis (top-right) and Rose wind charts of Ecosystem Services indicators (Bottom). The charts indicate normalized values of supply (right circle) and demand (left circle)

Land Use features, Dim2 has minor significance and cannot be related to specific municipal characteristics. The graph shows how municipalities included in C1 Urban Coastal and C5 Mountain Forest are noticeably separate from the rest, while C2, C3, and C4 have multiple intersections. Together with the spatial map and the visualization of ES values through wind rose charts, Fig. 3.5 presents the ES Bundles characterization in Le Marche regional case study.

3.2.3   Socio-Economic Characterization, Trade-Offs, and Sinergies

Methods for Socio-Economic Characterization

The socio-economic characterization complements the ES analysis by linking landscape patterns to human activities and demographic trends. It helps to interpret how social, economic, and land-use factors influence the configuration and performance of regional social-ecological systems. For the socio-economic characterization of social-ecological systems, three groups of indexes were chosen: Social, Economic, and Land Use indicators. Table 3.3 provides the socio-economic indicators selected and their definition. The spatial correlation analysis was carried out using R software Corrplot package (Taiyun and Viliam 2021). To identify weak and strong relationships (existence of synergies and trade-offs), associations between pairs of ES were detected using Pearson parametric correlation test.

The *Social indicators* aim at highlighting the main social components of territorial systems. The S1 Demographic index measures the percentage ratio between the population of elderly age (65 years and older) and the population of young age (less than 15 years). The S2 Social and material vulnerability index estimates the vulnerability for each territory, based on seven different dimensions of "material" and "social" vulnerability. The higher it is, the greater is the risk of discomfort in that area. The S3 Income accounts the municipal means of per capita income (euro/person).

The *Economic indicators* follow the three-sector model in economics, focusing on specific activities relevant for the study. E1 Primary sector index looks at the people employed in companies extracting raw materials, including the ATECO classes A—Agriculture, forestry and fishing and B—Extraction of minerals from quarries and mines. E2 Secondary sector index focuses on people employed in manufacturing companies, specifically the ATECO classes C—Manufacturing activities. Finally, E3 Tertiary sector considers employees from companies from the ATECO classes I—Accommodation and food service activities and R—Artistic, sporting, entertainment, and recreational activities.

The *Land Use indicators* aim to read the territory through the Corine land cover (CLC) data, calculated in municipal units through spatial analysis (QGIS). L1 artificial surfaces read the incidence of artificial surfaces (CLC layer 1) on the total area. L2 Agricultural surfaces looks at the incidence of agricultural surfaces (CLC layer 2) on the total area, L3 Forests and seminatural areas the incidence of Forests and seminatural areas (CLC layer 3) on the total area.

The socio-economic characterization was developed through the calculation of the mean value of each normalized indicator within the municipalities included in the Bundles. The results are then displayed in a bar graph according to the social, economic, and land use characteristics.

Table 3.3 Socio-economic indexes, relative unit, and definition. Abbreviations: "emp" stands for employed in the economy sector, "inh" stands for inhabitants of the municipality

Social indicator	Unit	Description
S1 demography (population aging)	K (age/age)	The indicator represents an index of population aging, consisting in the percentage ratio between the population of elderly age (65 years and older) and the population of young age (less than 15 years)
S2 social and material vulnerability	K	The indicator is structured through the combination of seven elementary indexes that describe the main "material" and "social" dimensions of vulnerability: (1) percentage incidence of the 25- to 64-year-old population without a degree; (2) percentage incidence of households with potential economic distress; (3) percentage incidence of households with potential welfare distress; (4) percentage incidence of population in severe housing distress; (5) percentage incidence of households with six or more members; (6) percentage incidence of single-parent young adult families; (7) percentage incidence of 15–29-year-olds who are inactive and not studying. The higher the value of the index, the higher the vulnerability level of the municipality
S3 income	Euro/inh/km^2	The indicator accounts the municipal average of total taxable per capita income. Unit: Euro/person.
Economic indicator	**Unit**	**Description**
E1 agriculture, forestry and fishing	Emp/1000 inh/ km^2	The indicator considers the municipal population employed in companies of primary sector. It accounts companies of ATECO classes A—Agriculture, forestry and fishing and B—Extraction of minerals from quarries and mines
E2 manufacturing activities	Emp/1000 inh/ km^2	The indicator considers the municipal population employed in companies of secondary sector. It accounts companies of ATECO class C—Manufacturing activities
E3 accommodation and recreational activities	Emp/1000 inh/ km^2	The indicator considers the municipal population employed in companies of secondary sector. It accounts companies of ATECO classes I—Accommodation and food service activities and R—Artistic, sporting, entertainment and recreational activities. Unit:
Land use indicator	**Unit**	**Description**
L1 artificial surfaces	Km^2 art/km^2 tot (%)	The indicator assesses the incidence of artificial surfaces (CLC layer 1) on the total area
L2 agricultural surfaces	Km^2 agr/km^2 tot (%)	The indicator assesses the incidence of agricultural surfaces (CLC layer 2) on the total area
L3 forests and seminatural areas	Km^2 nat/km^2 tot (%)	The indicator assesses the incidence of forests and seminatural areas (CLC layer 3) on the total area

Results: Socio-Economic Profiles

Figure 3.6 shows a summary of the spatial patterns of socio-economic indicators (the single maps can be found in the supplementary materials—Appendix B). Looking at social indicators, S1 Demographic index shows clear imbalances between inland areas and urban poles with younger residents mapped in the coastal municipalities and the inland systems at serious risk of depopulation. The map of S2 Social and material vulnerability displays various results across the region, with high values of the indicator mostly in the central southern part of the region. S3 highlights a clear link between income of population and the inland areas limits, with the inland municipalities presenting lowest regional per capita income values.

As for the economic indicators, the E1 reveals a significant prominence of agriculture, forestry, and fishing sector for the south of the region. The indicator E2 related to manufacturing activities presents the highest values for the sub-urban

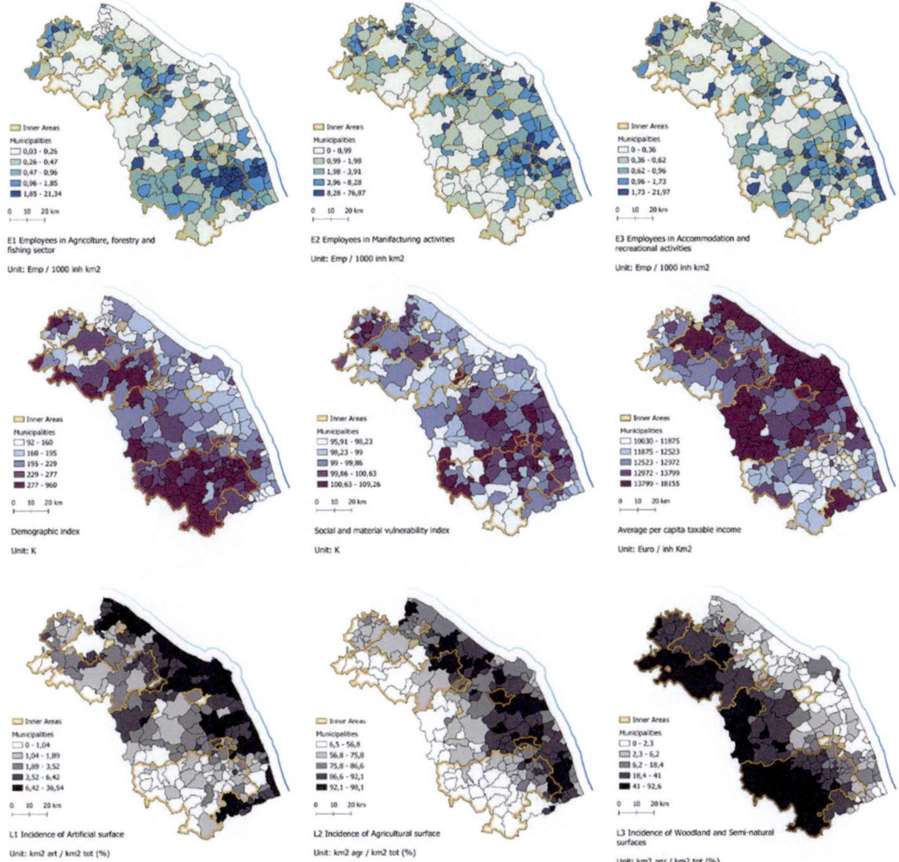

Fig. 3.6 Spatial patterns of socio-economic indicators. Single higher-definition maps can be found in the Appendix B

municipalities close to the urban coastal continuum. Finally, the accommodation and recreational activities (E3) highlight peak values in the coastal southern municipalities (probably connected to seaside tourism) and individual hotspots in the inner side of the region. It should be noted that the mapping presents in the case of economic indicators higher values for smaller municipalities.

As for the Land Use, L1 shows the incidence of artificial surfaces on the total municipal value and emphasizes a clear contrast between the coastal municipalities and the inner territories. The map shows artificial land use strips infiltrating from the coast toward the mountains especially in one central and one southern valley of the region. L2 highlights a central low-hill area, close to the seaside, dedicated to agriculture, in opposition to low—close to zero—values in the mountain municipalities. L3 shows a mountain belt of woodland and seminatural areas, with lower-altitude spots connected to protected areas closer to the coast.

Correlating Socio-Economic Indicators with ES Bundles

The correlation analysis shows that socio-economic indicators are mostly significant in characterizing the five detected bundles. Especially, among the social indicators, the most significant correlation is found in S1 Demography (cor = 0.45) and S3 Income (−0.31). The positive correlation between Bundles number (as defined from B1-B5) and aging index proves a tendency of aging of population along the coastal-mountain gradient, while the negative correlation of the income shows how wealth decreases toward mountain areas. On the other hand, S2 Vulnerability does not show significant correlation with the bundles structure (p-value >0.05). As for the Economic indicators, E1 Agriculture does not present a significant correlation, while E2 Manufacturing and E3 Accommodation and recreation have a less significant correlation (respectively cor = −0.14 and cor = −0.22). Finally, the Land Use indicators show correlation values over 0,6 and prove to be useful to characterize the ES bundles. The values of L1 Artificial areas characterize mostly the B1 Urban bundle (46% or land use), with lower values for all the other bundles. L2 Agricultural areas describes the best bundles B2 and B3 (respectively 91% and 90%), but also highlighting moderate values in B1 Urban areas (79%) and B4 Mosaic cropland forest (66%). Lastly, L3 Woodland and seminatural areas gives the highest values in B5 Mountain forest (70%), followed by B4 Mosaic cropland forest (34%). Figure 3.7 shows the socio-economic data according to the significance of the indicator to characterize the bundles.

The Pearson parametric test between pairs of ES indicators showed significant correlations among many ES in the case study of Le Marche region (Fig. 3.8). Especially, we observed a pattern of trade-offs, first between the supply of P1 Agricultural products and all the regulation services, together with C1 Eco-tourism and C3 Mushroom picking. Smaller negative correlation was also detected between those services and P2 Wine products. In terms of demand, light trade-offs can be noticed between the demand of R1 Hydraulic regulation and the provisioning services, and also between the demand of R1 Hydraulic regulation and R2 Soil protection and the cultural services and R4 Crop Pollination. Looking at synergies, strong positive correlation was found among the supply of all the Regulating services, together with C1 Eco-tourism and C3 Mushroom picking. In terms of demand, very

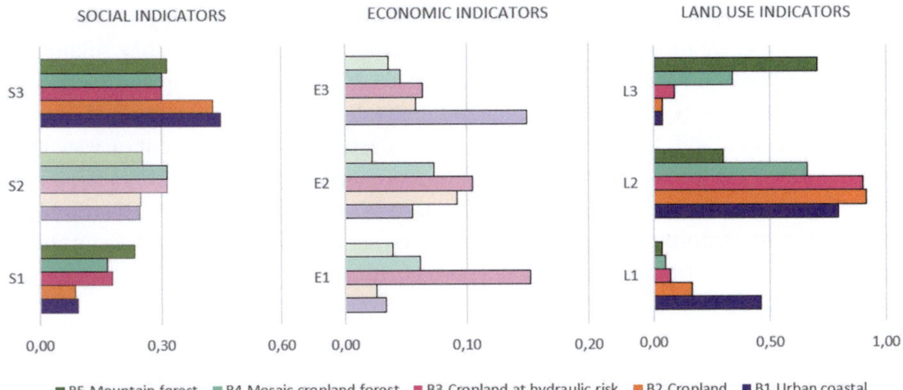

Fig. 3.7 Socio-economic characterization of the Ecosystem Services Bundles. The bars indicate average normalized values of the socio-economic indicators within the municipalities included in the Bundles. Transparency indicates low correlation between the indicator and the bundles categories

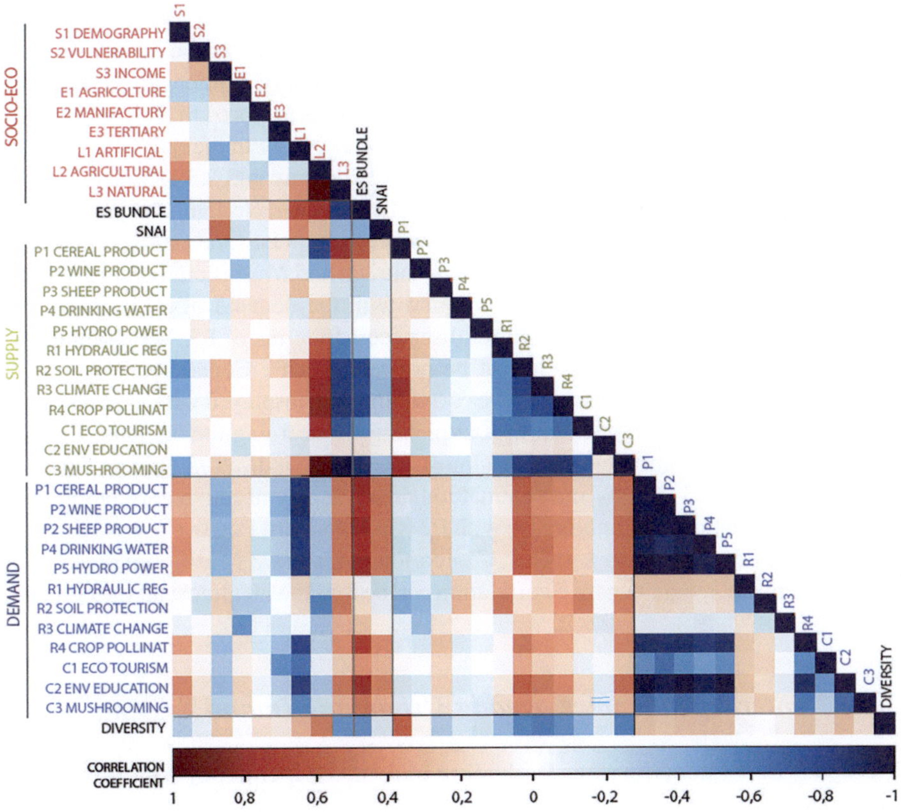

Fig. 3.8 Spatial correlation between ES demand indicators (left) and ES Supply indicators (right)

strong synergy was found between the demand of all the provisioning services, R4 Climate change regulation and C2 Environmental education, probably related to population density. Also, C1 Eco-tourism and C3 Mushroom picking showed a significant correlation with the urban demand areas. When comparing areas of supply with areas of demand, Fig. 3.8 shows that the correlation is often absent or very low, proving how the areas of demand differ from areas of supply. Regarding the Diversity index, a strongest negative correlation is shown in relation to P1, reinforcing the results denoting agricultural production as a limit factor for the supply of other services. On the other hand, a positive correlation is shown within the regulation services, C1 Eco-tourism and C3 Mushroom picking.

Identifying Trade-Offs and Synergies Among Ecosystem Services

To explore links between ES behavior and local characteristics, the correlation analysis dedicates a deeper look with respect to socio-economic characteristics. In terms of the S1 Demography index, referring to aging of the population, it is possible to observe: (1) a weak negative correlation with the supply of P1 cereal production, (2) weak positive correlations with the supply of all the regulating services and Eco-tourism and Mushroom picking, (3) a negative correlation with the demand of the majority of ES, and (4) a positive correlation with the Diversity in ES Supply. While with S2 there are no relevant correlations observed, the situation with S3 is mirrored to S1.

Economic indexes do not show noticeable associations, with the only exception of a minor positive correlation between E1 and the P2 supply and R3 Demand. On the other hand, land use indexes represent the main factors characterizing the bundles. L1 Artificial surfaces shows light negative correlation with P3 Pastoral products and very negative with all the regulation services, Eco-tourism and Mushroom picking. On the other hand, very positive correlation is found with the demand of all provisioning services, crop pollination, and the cultural services. In terms of L2 Agricultural surfaces, a strong positive correlation can be noticed with P1 Cereal production and, lighter, with P2 Wine products. Negative correlations are displayed with all the regulation services, eco-tourism and Mushroom picking. Concerning the demand, Fig. 3.7 shows light positive correlation with almost all the ES indicators (except for eco-tourism). Finally, L3 Forests and seminatural areas display negative correlation with P1 Cereal production, and lightly also with P2 wine products. Positive correlations are found with all the regulating services, C1 eco-tourism and C3 Mushroom picking. In terms of demand, a light negative correlation can be found with almost all the demand.

3.2.4   Interdependencies Among Urban and Inland Systems

Methods: Ecosystem Services Budgeting and Interdependency Analysis

The budgeting analysis was developed to explore how the balance between ES supply and demand reveals potential flows of goods and services across the region. This approach helps identify spatial dependencies among territorial systems and quantify surpluses or deficits of ecosystem benefits. To hypothesize flows of goods and services, the information on supply and demand of services was calculated in ES budgets as the difference between demand and supply (Burkhard et al. 2012). To visualize possible interdependencies among urban poles and inland areas, the budgets are summed within each category. To compare supply and demand indicators we reclassified in a 1–100 scales the ES R1 Hydraulic regulation, R3 Crop pollination, C1 Eco-Tourism, C2 Environmental education, and C3 Mushroom picking.

Results: Territorial Interdependencies and Ecosystem Service Budgets

A strong correspondence was found between the composition of Bundles with respect to the SNAI classification (Fig. 3.9). This is particularly clear for the Bundle B1 which includes the main urban and inter-municipal poles, together with belt areas. Similarly, B5 and B4 are mostly dominated by outermost and peripheral inland areas. Focusing on the Bundles gradient from the coast to the mountains, the graph shows a general trend of growing number of inland municipalities and declining of belt areas. Nevertheless, it is interesting to notice how the urban poles are still present in the Bundle B5. This result draws the picture of a polycentric territory, with major settlements also part of the mountain bundles. Lastly, noteworthy is the absence of urban poles from B3, characterizing the Bundle as lower population density.

Figure 3.10 shows the supply–demand budget within the SNAI municipal categories. It demonstrates the extent to which each category has a deficit or a surplus within Provisioning, regulating, and cultural services.

Starting from the Provisioning services, it is possible to observe how the budget of ES P1, P2, and P3 for Urban Poles gives a deficit, with a peak on P2 (−0,45) and P3 (−0,56). Similarly, inter-municipal poles present minus values except for a net value for P1. On the other hand, Inland areas offers a surplus for all the values, reaching the maximum in the ES P1 (+0,41). Looking at the water resources, the

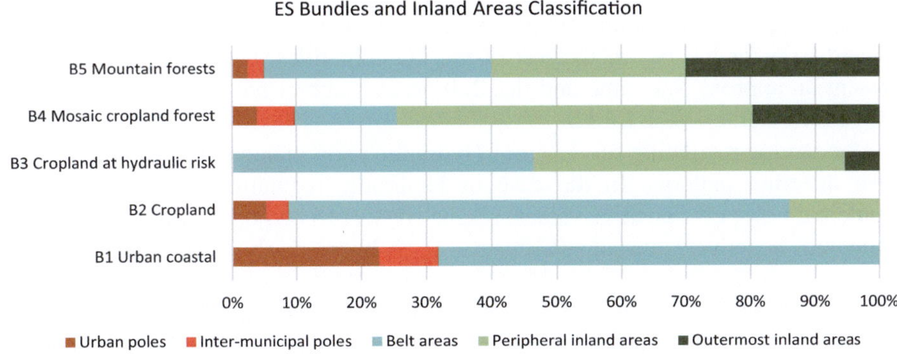

Fig. 3.9 Characterization of Ecosystem Services Bundles in respect of the SNAI Inland Areas classification

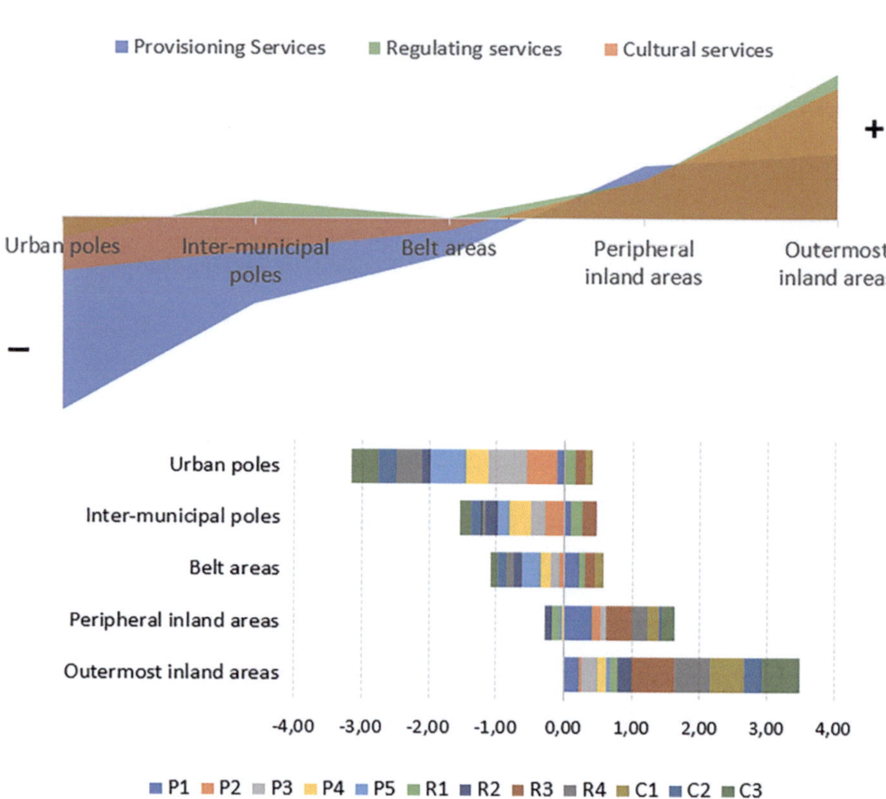

Fig. 3.10 Area chart (up) of the Ecosystem Services categories Budget for each SNAI municipal category, and bar chart (bottom) summing up deficit and surplus per each Ecosystem Service

picture of dependency of Poles toward inland areas becomes lighter: the strong deficit presented by the Urban poles (−0,34 for P4 and − 0,52 for P5) is not matched by a strong supply by the Peripheral inland areas, which have a surplus of only 0,14 for P4 and 0,06 for P5. As for the rest of the municipalities, Peripheral inland areas present an approx. Net value and the Belt areas a lighter but still relevant deficit: −0,16 for P4 and − 0,26 for P5.

Concerning regulating services, R1 Hydraulic regulation and R2 Soil protection have differing patterns. In the case of Hydraulic regulation, urban and inter-municipal poles give positive values, respectively, 0,15 and 0,19, while soil protection yields negative values of −0,12 and − 0,19. Interestingly, the peripheral inland areas present deficits both for R1 (−0,09) and R2 (−0,12). Regarding services related to biological cycles, the R3 Pollination shows a positive balance for all the municipal categories, whereas the Climate Change regulation (R4) reveals another imbalance, with large deficits connected to urban poles and surpluses in budget within inland municipalities.

Finally, cultural services are considered. Surprisingly, C1 Eco-tourism presents a positive budget for all the municipal categories (except a net value for Inter-municipal poles), with the peak reached by Outermost inland areas reaching a value of +0,50. On the other hand, C2 Education and C3 Mushroom picking show strong deficit for urban poles (respectively −0,27 and − 0,40) and surplus values in Outermost inland areas (+0,27 and + 0,56). However, as supply and demand values for cultural services were reclassified from different units it is important to note that the comparison has no values in absolute numbers but only in spatial terms.

3.2.5 Patterns of Social-Ecological Interaction Using ES Bundles

The study proved useful in characterizing landscapes as social-ecological systems relating ES patterns to local socio-economic characteristics. In respect to other researches applying the concept of Bundles in the relation between demand and supply of ES (Baró et al. 2017; Peng et al. 2020; Queiroz et al. 2015; Quintas-Soriano et al. 2019), the present work integrated the ES analysis with a socio-economic characterization, allowing the interpretation of patterns of human–nature interactions. These patterns are the result of complex interactions associated with local, historical, and cultural processes (Antrop 2005), which in Le Marche as in other Mediterranean regions, are shaped by the combination of the ecological and biophysical assets, together with local management practices (Balzan et al. 2018; Blondel 2006). The Bundles analysis identified five landscape systems based on 12 ES supply and demand which were then characterized through nine socio-economic indicators.

The regional landscape systems can be interpreted along a coastal-mountain gradient drawn by the raising of altitude and decreasing of population density. Further characterized in other studies through land cover characteristics (e.g., agricultural land cover in Queiroz et al. 2015 and Quintas-Soriano et al. 2019), here the urban-inland classification highlights a polycentric regional feature that locates major urban poles also in mountain Bundles. This validates the limitation of the urban–rural dichotomy beyond metropolitan areas, where local characteristics are not captured through regional-scale analysis (Grêt-Regamey et al. 2014). In our study, the urban Bundle includes coastal areas and small urbanized municipalities along the valleys, excluding major inland urban centers characterized by lower population density. Hilly and valley municipalities which don't fall in the urban system are part of Bundles 2 and 3, named "Cropland" and "Cropland at hydraulic risk." In this context, several studies show how maximizing agricultural production is likely to have effects in the decrease of regulatory services (Balzan et al. 2020; Felipe-Lucia et al. 2020). As regulating services are rarely the main emphasis of trade-offs, but often are the most impacted (Turkelboom et al. 2018), it is important to connect the socio-economic patterns with the provision of ES. The regional case study presents a picture of small-scale and diverse farming systems in synergy with local ecosystems (Bevilacqua 2013), nevertheless, the analysis indicates a relevant trade-off between agricultural production and the set of regulating ES.

Following along the coastal-mountain gradient, the bundles 4 and 5 ("Mosaic cropland forest" and "Mountain Forest") correspond to those landscapes hosting a share of woodland and seminatural areas. While bundle 4 is still characterized by minor agricultural activities (and consequent ES supply of cereal production), bundle 5 shows high levels of all services related to forested land use (Balzan et al. 2020). Confirming findings advanced by other studies (e.g., Baró et al. 2017; Felipe-Lucia et al. 2018; Raudsepp-Hearne et al. 2010), a strong synergy is displayed in forested mountain areas between regulating and recreational services. In these systems, the ecological benefits associated with afforestation are linked to recreational opportunities and esthetics appreciation, which are also associated with the presence of built infrastructure and human activities (Langemeyer et al. 2018).

> **Box 3.2 Interpreting Landscapes as Social-Ecological Systems through ES Bundles**
>
> The classification of municipalities into ES bundles offers a practical lens to interpret landscapes as integrated social-ecological systems. While traditional bundling approaches (e.g., Raudsepp-Hearne et al. 2010) focus solely on ES indicators, this study expands the perspective by integrating socio-economic variables, thus enabling a more systemic interpretation of landscape dynamics.
>
> By grouping spatial units based on similarities in ES supply and demand, and then characterizing them according to socio-economic attributes, the resulting bundles reflect differentiated functional roles within the region—ranging from high-pressure urban areas to multifunctional rural landscapes. These clusters reveal spatial gradients, trade-offs, and interdependencies that are not captured by conventional land use classifications alone.
>
> This extended bundling approach could serve as a basis for new forms of landscape characterization, particularly relevant for regional planning and policy. In this regard, it may offer a valuable contribution to the elaboration of Regional Landscape Plans, where ecological functionality and social values are increasingly recognized as foundational components of landscape quality.

3.2.6   Implications for Sustainable Landscape Development

This study offers key elements for the sustainable development of landscapes as the information on ES patterns is proved to be useful for landscape planning by many practitioners (Albert et al. 2014; Mascarenhas et al. 2014). The added value lies in improved opportunities for integrating local assets in management measures, accounting trade-offs and synergies within local stakeholders and developing targeted response measures (Albert et al. 2016). We highlight in this section three key elements for regional landscape planning related respectively to (1) the preservation of the identity of inland systems by supporting local ecosystem management and responsible ecotourism strategies, (2) the enhancement of sustainable agriculture through small-scale farming and sustainable practices (also in relation to CAP), (3) a special attention on multifunctional landscape management practices, such as pastoralism.

The first regards the management of inland systems under economic and demographic instability. From a socio-economic perspective, the inland systems are characterized in our study by a high aging rate and low incomes. Those trends are confirmed in other Mediterranean regions (Balzan et al. 2020) and poses major challenges to regional planning for their impact in land management (Bruno et al. 2021). While abandonment in remote ecosystems carries along reforestation and a consequent increase in regulatory services, such as water regulation and soil retention, it is also important to ensure forest management to increase structural heterogeneity and therefore the supply of multiple ES (Felipe-Lucia et al. 2018). Traditional management practices can be promoted through the enhancement of eco-tourism strategies, based on a fruitful interaction with local systems (Aretano et al. 2013). In our mapping for Le Marche Region, high supply values in recreational services may indicate a potential development also confirmed by national strategies (MIBACT 2017). However, tourism development must take into account possible treats for the cultural identity of an area, together with possible degradation of natural systems due to mass tourism and the impact of the relative infrastructure (Gössling 2002).

The second element regards the sustainable management of agricultural land. As we known, the maximization of provisioning services also leads to changes in ecosystem functioning and biodiversity loss (Felipe-Lucia et al. 2020). These trends are partly mitigated in Le Marche regional case study by the growing rates of organic production (SINAB 2020) and the small-scale system which characterizes local agriculture (Bevilacqua 2013). This latter factor is increasingly addressed by the literature as a crucial factor for supporting biodiversity. Tscharntke et al. (2021) argue that increases in cropland heterogeneity with at least 20% seminatural habitat per landscape should be a key recommendation in current biodiversity frameworks. This topic is today at the center of a major political debate as the common agricultural policy (CAP) 2021–27 is being discussed. Despite the poor expected positive benefits in terms of environmental protection and climate change mitigation (Pe'er et al. 2019), the new CAP embraces eco-schemes based on the needs and priorities identified at national/regional level. In this sense, regional governance can move in the direction of integrating local assets into new programming.

Finally, we underline a third element connected to landscape multifunctionality, considered crucial for biodiversity conservation and human well-being (Balzan et al. 2020). Although mountain landscapes have the greatest ES supply, the regional case study associates foothill landscapes with the greatest diversity rate. In addition to conservation approaches, which are already implemented in the region, practices that balance agricultural productivity with social-ecological benefits should be promoted as agri-environmental measures (Iniesta-Arandia et al. 2015). This is the case of pastoralism, which in our study relates to the inland systems. This practice is considered a main factor for shaping cultural landscapes, as well as a practice of biodiversity protection (Oteros-Rozas et al. 2014). Globally declining (Dong et al. 2011), pastoral systems are considered to be vulnerable and should be supported for the variety of services they provide, including food security in a climate change (Krätli et al. 2013).

3.2.7 Implications for Regional Cohesion

Cohesion strategies aiming at equitable and just regional development must build on existing territorial links and interdependencies. The study showed how urban development is strictly based on its linkage with inland areas on which it depends for the availability of natural resources (Gebre and Gebremedhin 2019). This dependency concerns all the 12 ES taken into account and leads to relevant consequences in terms of environmental equity (Bennett et al. 2015; Pascual et al. 2014).

Among the main results, the study associates the lowest income to the main hotspots of ES supply. Those partially coincide with inland areas which are characterized by an income gap with urban poles (Romagnoli and Mastronardi 2020). To address this gap from an ES perspective, a growing number of policies apply the concept of Payment for Ecosystem Services (PES) to promote environmental conservation and social development goals (Wunder et al. 2020). However, the PES strategies are often designed on a single environmental objective (e.g., compensation for the delivery of fresh water form coastal to mountain municipalities), and single-objective approaches are proved to show limitation in fostering fair and balanced development (Benra et al. 2022). Pascual et al. (2014) stress how the support should go beyond the distribution of income or benefits but rather take into consideration the different dimensions of equity. Those also involve procedural dimensions including the role of local actors in decision-making and their cultural identities, values and knowledge systems (Pascual et al. 2014).

In the context of regional development, place-based strategies aim at integrating the local assets and knowledge in the design and delivery of public policies (Barca et al. 2012). It is opposed to spatially blind provision of public subsidies that have characterized the past welfare policies for inland areas in Italy (Viesti 2016). Though place-neutral policies, public subsidies have led to the sole outcome of reducing social tensions, weakening the political weight of inland areas (Barca 2018). The ES perspective, on the other hand, can support the characterization of local assets emphasizing the role and perspective of local actors in landscape management (Felipe-Lucia et al. 2015). Especially in Mediterranean areas, integrated methodologies and stakeholders' involvement have been scarcely applied (Nieto-Romero et al. 2014) and it is crucial that cohesion policies support participatory processes when defining interventions. It is especially relevant as responses to trade-offs depend on the level of awareness of stakeholders (Turkelboom et al. 2018) and the civic engagement itself constitutes social-ecological process that directly generates ES and benefits for human well-being (Krasny et al. 2014).

The regional cohesion can be further analyzed through the actual flow between areas of supply and demand classifying ES according to their proximity and suitability of use (Burkhard et al. 2014). While the food services are considered "*decoupled*" as they can be transported and imported from elsewhere, others, such as fresh water, hydraulic risk, and soil protection can be seen as *proximal*, as the user systems need to be geographically related to providing areas (Baró et al. 2017). This poses a certain urgency in the management of proximity services. The regional case study shows how urban areas are dependent on inland systems for the provision of regulating services, where changes in supply systems might have direct effects on the land stability of the regional systems.

> **Box 3.3 Telecoupling and Power Relations: Avenues for Future Research**
> A growing field of research investigates the effects that ES demand in one place can have in the socio-ecological dynamics of a distant place (Sonderegger et al. 2020). This field is addressed within the definition of "telecoupling", which takes into account cross-scale social relations, as decisions made at local scales are often shaped by actors at larger scales (Martín-López et al. 2019). Its integration into the regional study could lead to interesting insights about environmental equity beyond the local dimension.
>
> Within regional boundaries, the service flow analysis could be further investigated to highlight stakeholders' power relations. Aspects such as land stewardship, access rights, and governance systems are stressed to be important in the definition of relationships between supply and demand for ES (Felipe-Lucia et al. 2015). Furthermore, Mapping Cultural ES through social analysis (questionnaires, participatory mapping, focus groups), can partly compensate the simplification given by indicators in assessing social-ecological phenomena (Baró et al. 2017).
>
> This study represents a first stone for the construction of Le Marche regional Green Infrastructure. Data on ES supply can support the establishment of robust decision support tools to facilitate decision making in land use management, to balance food production from agriculture sector as well as environmental protection (Morri and Santolini 2022). A comparison of the ES Bundles with the existing regional ecological networks (REM) could further investigate the topic of biodiversity and ecological connectivity.

3.3 Local Perceptions and Co-Production

A key challenge in regional planning lies on the integration of social components in environmental evaluation, in order to take into account the complexity of local social-ecological systems. Those systems are the result of the action and interaction between nature and human activity, which shaped the territory to produce food, fiber, timber, mostly consisting of private goods, while natural systems provided other services, predominantly public, crucial for society (Früh-Müller et al. 2016). The last decade has seen increasing attempts to assess this interaction through the concept of landscape, considered the most suitable spatial unit for managing ecosystems (Forman and Godron 1986; Tallis et al. 2015) and whose definition allows the valuation of the physical entities and their perception (European Landscape Convention, 2000). In this context, landscape planning faces the challenge of reconciling competing sectorial interests working on the same areas for different purposes, in order to guarantee the multifunctionality of landscapes as a condition for a sustainable development (de Groot et al. 2010a, b; Sargolini and Gambino 2016).

The ES framework is increasingly applied by planners and decision-makers for its capability to integrate in one assessment a great variety of benefits humans derive from ecosystems (IPBES 2019). Several studies developed tools for the valuation of provisioning, regulating and cultural services in order to support planning recognizing spatial environmental relations (Albert et al. 2016; Grêt-Regamey et al. 2017; Langemeyer et al. 2016). These frameworks mostly refer to the assessment of physical components, to support land-use decision making toward the goals of sustainable development (Grêt-Regamey et al. 2017). Yet, development strategies need to integrate the social dimension of sustainability to support resilience of communities through their role in the management of ecosystems (Bennett et al. 2015).

Today, a growing body of literature is focusing on the concept of ES co-production (Fischer and Eastwood 2016; Jericó-Daminello et al. 2021; Lavorel et al. 2020; Palomo et al. 2016), stressing how benefits from nature to people does not occur independently but in most of the cases require a significant human contribution (Díaz et al. 2015). Co-production of ES includes several anthropogenic components relating with natural systems, such as motivation or education (human capitals), values and norms (social capitals), machinery and infrastructure (physical capital), and credits or direct payments (financial capitals) (Palomo et al. 2016). ES can be co-produced by anthropic activities through the direct management of service flow (e.g., agricultural activity, forest management) or as users, benefiting and valuing a service (e.g., the preference for a product or a tourist destination, or the ecosystem function toward an environmental risk). Together with stakeholders interested in or investigators on the services, those roles shape the interaction between society and natural systems and are crucial in the investigation on landscape as social-ecological systems.

Figure 3.11 illustrates the theoretical framework proposed in this study, distinguishing physical from cognitive co-production (Fischer and Eastwood 2016; Palomo et al. 2016). The former refers to the physical action on ecosystems involving measurable external changes and relates to the anthropogenic and natural contribution on the landscape. The latter belongs to the cognitive processes related to the individual perception of the benefits and addresses the importance of perceived

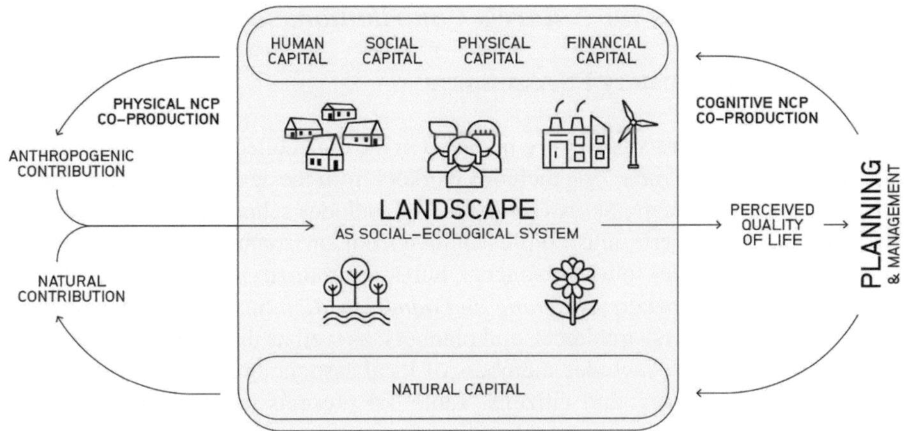

Fig. 3.11 Ecosystem Services co-production in the landscape as social-ecological system

quality of life for landscape planning and management. The application of the ES co-production framework on landscapes as social-ecological systems has the potential to assess dependencies and benefits of stakeholders for ES and to highlight collaborations among them (Opdam et al. 2015a; Turkelboom et al. 2018). The acknowledgment of the role of actors and its implication on stakeholders relations can provide planning with an integrated valuation tool, including stakeholders perspectives in the management of ecosystems (Rieb et al. 2017).

However, the contribution of social systems to ES co-production remains unclear. There is a lack of evidence on how people value ES and perceive benefits to their quality of life, according to their role in society (Bennett et al. 2015). It is not clear how different stakeholders depend on and benefits by which ES, and at which scale they interact with each other (Martín-López et al. 2019). In this frame, the analysis of spatial relations is crucial for understanding power asymmetries and the distribution of ES among the beneficiaries (Bennett et al. 2015). Furthermore, recent findings (Jericó-Daminello et al. 2021) highlight discrepancies between how stakeholders perceive themselves as co-producers and how others perceive them.

Against this background, the present study[3] aims to test the integration of perception of social actors in landscape planning through their role in ES co-production. Co-production is used as a framework to explore local perception of different stakeholder groups and understand anthropic contributions to ES offered by landscapes. Following an approach that explicitly refers to the planning dimension, we look at (a) the main ES perceived by local stakeholders, (b) the anthropogenic contributions involved in ES co-production, (c) the implications of ES co-production on the relationship among stakeholders.

---

[3] Part of this study (Sect. 3.3) was originally published as the article "Including the perspective of stakeholders in landscape planning through the Ecosystem Services co-production framework: an empirical exploration in Le Marche, Italy," by Matteo Giacomelli, Massimo Sargolini, and María R. Felipe-Lucia, in Regional Environmental Change (2024) 24: 2184, https://doi.org/10.1007/s10113-024-02184-w

3.3.1   Preferences for Nature's Contributions to People

Methods for Participatory ES Assessment

Social actors of Fiastra Valley were involved in the data collection according to five stakeholder groups: *Production* includes workers from the agriculture sector, agronomy, and local producers; *School and research* includes school teachers, university students, ecology experts, and a representative from environmental centers; *Culture and commerce* includes tourist managers, hotel, agritourism and restaurant owners, café and local shop owners; *Planning and administration* includes members of the municipality, engineers, architects, and planners, as well as the local water distribution company; *Society* includes members of local associations, artists, family doctors, local recreationists, and citizens. Table 3.4 presents the stakeholder groups addressed in the study.

To identify and assess the preferences stakeholders relate to local ES in July 2021, we organized focus groups involving a total of 27 people. To facilitate the activities, participants were divided in two groups, according to their roles:

- Focus group A—Civil Society, composed by *School and research* (school teachers), *Society* (members of associations), *Production* (farmers and local producers) (15 participants).
- Focus group B—Management, composed by *Planning and administration* (municipal representatives, planners, and architects) and *Tourism and commerce* (tourist managers) (12 participants).

After introducing the concept of ES, the participants were asked to take part in a free listing exercise, answering questions about the benefits that society in the Fiastra Valley receives from ecosystems. Participants had 15 min to answer in private, with a maximum of three post-its per question: (1) What does nature in the Fiastra Valley mean to you? what is the importance of nature for this area? (2) What are the benefits that Fiastra Valley provides for your well-being/to fulfill your organizational goals? (3) What goods and products does nature in the Fiastra Valley provide to society? How does nature support the local economy?

The post-its were collected and analyzed by facilitators which extrapolated individual concepts addressing natural contribution to people. Those concepts were finally grouped following the ES classification. The final list of ES was displayed on a board and participants were asked to vote the three ES they considered most relevant for the Fiastra Valley.

Results: Stakeholder Preferences for Ecosystem Services

Participants of the Focus Groups indicated, through free listing, a great variety of benefits society receives from Ecosystems in the Fiastra Valley, resulting in 147 statements related to ES. As given in Table 3.5, where answers are grouped in ES categories, most of participants included statements which were grouped as *C7 Mental well-being*, and many also addressed *C5 Sense of Place* (24 participants) and *C1 Eco-Tourism* (16 participants). Only participants of the Focus group 1 (Civil Society) addressed *C2 Environmental education* and only participants from Focus group 2 (Management and administration) addressed *P2 Drinking water*. *R2 Climate regulation* was addressed only by one statement from Focus group 1.

Furthermore, the participants of the focus groups voted their preferences for the defined ES and showed a high preference for cultural ES (Fig. 3.12). The most voted ES were: *C6 Esthetic beauty* (19% of votes), *C5 Sense of place* (16% of votes), *C8*

Table 3.4 Stakeholders of Fiastra Valley Ecosystem Services, grouped in stakeholder groups and participants to the local questionnaire

Stakeholders Groups	Stakeholders	Scale	Description	Questionnaire participants
Production	Agriculture	Local	Actors involved in the agriculture sector, both farmers and agronomists	5
	Local producer	Local	Actors involved in artisanal activities made in a traditional or non-mechanized way using local natural resources	2
School and research	School	Local	Teachers from schools of the Fiastra Valley, including primary and first-grade secondary schools.	3
	Environmental center	Local	Representatives from local centers for environmental education	1
	Ecology expert	Regional	Academics and researchers from universities or regional ecologists and sociologists	1
	University students	Local	Students enrolled in university living or having roots in the Fiastra Valley	2
Tourism and commerce	Tourist manager	Local	Managers of local tourism, including travel agencies, tour guides, mountain guides.	2
	Regional tourists	Regional	Tourists from Le Marche region	–
	Foreigner tourists	Inter-regional	Tourists from out of Le Marche region	–
	Nearby distribution	Local	Grocery stores with local products	1
	Large distribution	Inter-regional	Supermarkets and inter-regional distribution chains	–
	Bars, restaurants and agritourisms	Local	Businesses where meals and drinks are served and places of aggregation for local population and tourists.	3
Planning and administration	Municipalities	Local	Fiastra Valley municipal authorities.	4

(continued)

Table 3.4 (continued)

Stakeholders Groups	Stakeholders	Scale	Description	Questionnaire participants
	Engineers, architects, and planners	Regional	Single people or studios involved in the technical environmental planning and management	3
	Regional authorities	Regional	Le Marche regional authorities	–
	Water distribution company	Regional	Representative from the local water distribution company	1
Society	Fiastra Valley citizens	Local	People living in the territory of the Fiastra Valley	3
	Urban citizens	Regional	People living in the main cities of Le Marche region	–
	Artist	Local	Local people who habitually practice creative productions in relation with nature	1
	Family doctor	Local	Local community doctor who treats patients with minor or chronic illnesses	1
	Associations for local promotion	Local	Cultural and environmental associations active in the local context	2
	Local recreationists	Local	Local citizens practicing sport or other open-air activities in the valley	1

Table 3.5 Nr. of statements related to Ecosystem Services expressed by participants at the Focus Groups during free listing activity

FG1	FG2	Tot	ES Group	FG1	FG2	Tot	ES Group
15	13	28	C7 mental Well-being	2	2	4	C3 sport activity
15	9	24	C5 sense of place	2	1	3	R3 air purification
10	6	16	C1 eco-tourism	0	2	2	P2 drinking water
7	4	11	P1 agricultural products	1	0	1	R2 climate regulation
5	6	11	C8 artisan products	0	0	0	P3 hydroelectric energy
7	3	10	C6 esthetic beauty	0	0	0	R1 hydraulic regulation
5	4	9	C4 intrinsic value				
5	0	5	C2 environmental education				

Artisan products (15% of votes), *C7 Mental Well-being* (10%). Four other ES were voted by 6% of the participants (*P1 Agricultural products, P2 Drinking water, C1 Eco-Tourism, C4 Intrinsic value*) and the rest received less than 5% of the votes. Nobody expressed a preference for *P3 Hydroelectric energy*. Looking at ES categories, besides cultural services being the most voted with 7,8 votes per service, provisioning services received an average of 3,3 votes per ES and regulating services got 2.

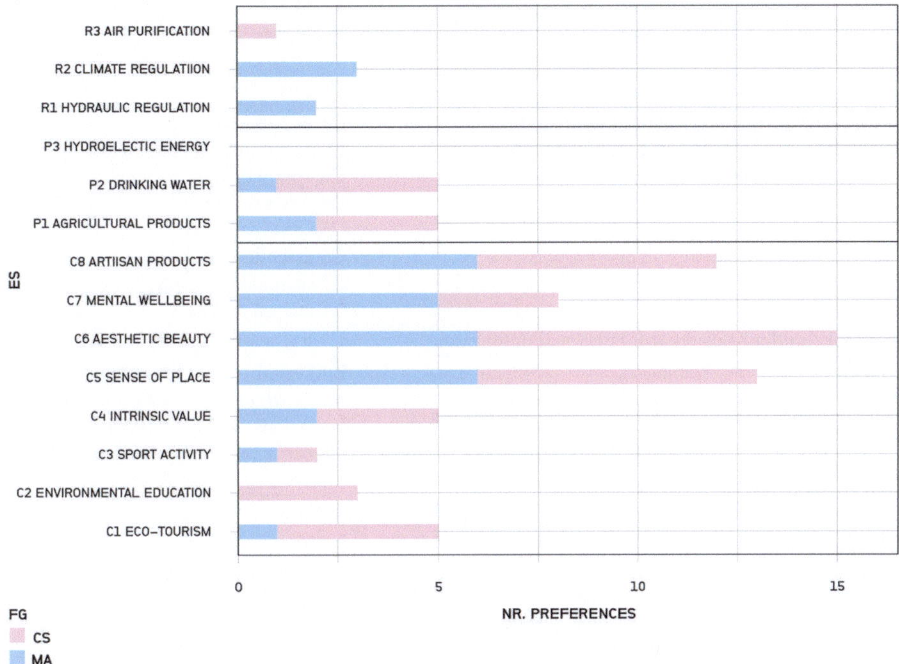

Fig. 3.12 Preferences for Ecosystem Services in the local case study. Data in red are from Focus group 1—Civil Society (CS), composed by: School and research (school teachers), Society (members of associations), Production (farmers and local producers). Data in blue are from Focus group 2—Management (MA), composed by Municipal and administration (municipal representatives, planners, and architects) and Tourism and commerce (tourist managers)

Looking at differences between the two focus groups, the statistical analysis stated that FG1 responses are not significantly different from FG2 responses. In addition, it is interesting to notice that *C2 Environmental Education* (4%) and *R3 Air purification* (1%) were considered only by participants from FG1, while *R2 Climate regulation* (4%) and *R1 hydraulic regulation* (3%) were chosen only by participants of FG2.

3.3.2 Anthropogenic Contribution in Co-Production

Methods for Questionnaire Design

To assess the role of local stakeholders regarding the co-production of ES, the participants of the focus groups were invited to respond to a questionnaire. The number of participants was integrated through snow-balling technique aimed at identifying additional respondents within the stakeholders identified in Table 3.6.

Table 3.6 Characterization of ES co-production, listed according to ES classification. (a) List of Ecosystem Services considered, (b) definition of roles for local actors, (c) capitals involved, (d) stakeholder involvement, (e) stakeholder collaboration

(a) Ecosystem Services List		
Provisioning ES	Regulating ES	Cultural ES
P1 agricultural products P2 drinking water P3 hydroelectric energ	R1 hydraulic regulation R2 climate regulation R3 air purification	C1 eco-tourism C2 environmental education C3 sport activity C4 intrinsic value C5 sense of place C6 esthetic beauty C7 mental Well-being C8 artisan products

(b) Stakeholder roles regarding ES and definitions				
User *Receive the benefits of the services*	Managers *His/her/their activity supports the offer of this service*	Negatively influenced *Bothered by the presence of the service*	Interested *Not a direct user but believe in its importance*	Investigator *You care about this service from a research point of view*

(c) Anthropogenic capitals involved in ES co-production and definitions			
Human capital *Knowledge, education, motivation, skills, or health*	Social capital *Values and norms, formal and informal networks, or trust*	Physical capital *Machinery, tools, infrastructure, or built capital*	Financial capital *Savings, credits, grants or direct payments*

(d) Stakeholder involvement		
Dependency for the ES *Which stakeholder rely on/depend on the service*	Benefit from the ES *Which stakeholder receives benefits from the ES*	Access to decision making *Who contribute to decisions related to the management or status of the ES*

(e) Stakeholder collaboration (*only for managers*)
Stakeholders with whom they would collaborate with, to provide the service

Overall, a total of 35 people took part in the questionnaire. The stakeholder group *Production* (7 participants) included workers from the agriculture sector (5), and local producers (2). *School and research* (8 participants) involved schoolteachers (3), university students (2), an ecology expert (1) and a representative from an environmental center (1). *Tourism and commerce* (6 participants) included tourist managers (2), Bar-, restaurant-, and agritourism- owners (3) and a local shop owner (1). *Planning and administration* (8 participants) included members of the municipalities (4), Engineers, architects, and planners (3), and a member of the water distribution company (1). *Society* (8 participants) included members of associations for local promotion (2), an artist (1), a family doctor (1), a local recreationist (1), and citizens (3).

To facilitate the compilation of the questionnaire, the respondents were assisted though face-to-face interviews. Due to safety precautions during the COVID-19 pandemic, interviews were mainly conducted by video-call, except in circumstances where respondents did not have access to online platforms. In case of qualitative information useful to justify eventual choices and support discussion, the results of the questionnaire were combined with notes taken by the interviewers. The interview lasted between 30 and 70 minutes per participant, depending on the number of services they expressed to play a role in, as structured in the questionnaire (see the full script in the Appendix C), following the five sections:

1. *Role in ES co-production.* The participants were asked to select one ES in which they play a role from the list of ES arranged by the focus group (Table 3.6 a), and specify which role they play in it (among the list Table 3.6 b).
2. *Capitals involved.* Participants were asked to assess non-natural capital inputs through which they are contributing to the ES co-production, namely: human capital, social capital, physical capital, and financial capital.
3. *Dependency, benefits, and influence on decision making processes.* From a list of stakeholders, the participants are asked to assess dependency on and benefits from the ES per each stakeholder. The same question is repeated for themselves. The participants are also asked to rate the access in decision-making both for themselves and the others.
4. *Collaboration among stakeholders.* In case the participants have a role as *manager* of the service, they are additionally asked to indicate which of the listed stakeholders they work with in the provision of the service.
5. The last part is dedicated to the *social-ecological characterization of the participants.* Here the participants are asked about their environmental behavior (frequency of visits to protected areas, consumption of organic and local products, waste recycling habits) as well as age, gender, occupation/activity.

Results: Stakeholders' Role and Capitals involved

Information on the role of stakeholder in ES co-production was extracted from the questionnaire sections A and B, that resulted in a total of 76 ES co-production records, representing in average 2,2 responses per person (Fig. 3.13). Respondents within the Stakeholder group *Production* expressed 9 ES records; *School and research* 18 ES records; *Tourism and commerce* 10 ES records; *Planning and Administration* 15 records; and *Society* 24 records. Respondents were aged between 22 and 71, with a balanced proportion of men and women. In terms of spatial distribution, they well represented the six municipalities of the case study (see Appendix F for details).

As a first result, it can be noticed how the local stakeholders stated co-production in all the ES of the list, with the exception of P3 Hydroelectric energy and R3 Air purification. The main ES to be selected were C5 Sense of place (13%), C1 Ecotourism (12%), C8 Artisan products (12%), and P1 Agricultural products (11%). All the others interested a share of 6%–9% of ES records, with exception of R1 Hydraulic regulation and R2 Climate regulation which regarded only 3% (2 ES records).

Respondents described their role in ES co-production mostly as Users (50,0% of ES records) and Managers (40,8%). Only 6,6% of records regarded the role

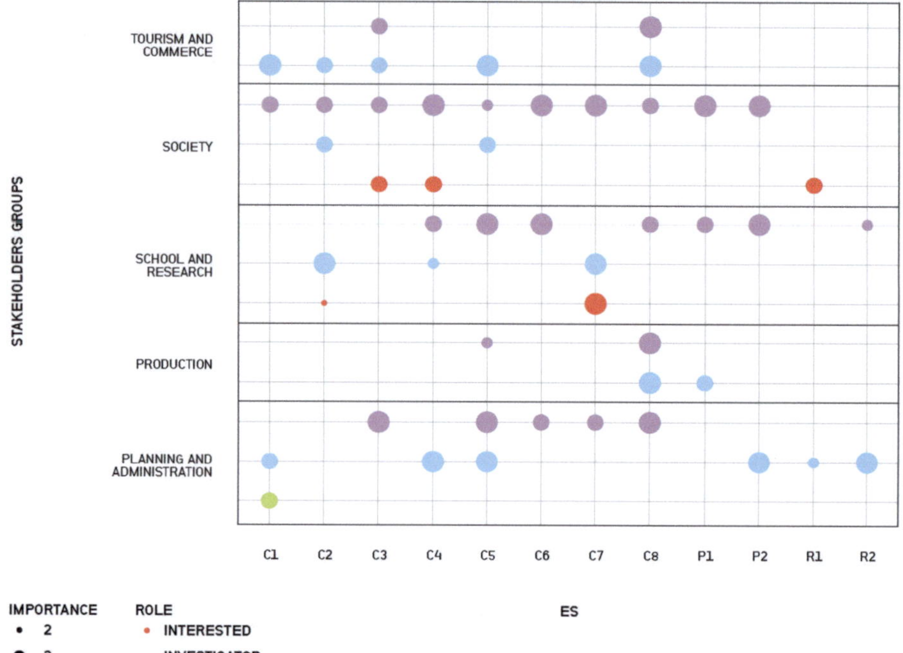

Fig. 3.13 Role of Stakeholders in Ecosystem Services co-production of the local case study. Importance was rated through a 1–5 Likert scale, the value 1 does not appear as no respondent selected it. Note that P3 and R3 are not represented in the graph as none of the respondents selected as having a role in them

Interested and 1,3% Investigator. The figure shows the results per Stakeholder groups, and differences can be noticed between the ES: looking at C5 Sense of place, 60% of ES roles account Users and 40% Managers; as regards C1 Eco-tourism 22,2% are Users, 66,6% Managers and 11,1% Investigators; C8 Artisan products includes 66,6% Users and 33,3% Managers; while P1 Agricultural products comprises 62,5% Managers and 37,5% Users.

Looking at the distribution of roles per Stakeholder groups, we see how the *Tourism and commerce* and *Production* groups refer to high rates of Managers, concerning, respectively, 80% and 78% of the ES records. The group *School and research* accounts 28% of Managers, 71% of Users, and 11% of Interested, while *Planning and administration* share an equal composition of Managers and Users (47%) with a 7% of Investigators. *Society* includes 75% of Users, 13% of Managers, and 13% of Interested.

As for the topic of capitals involved, most of respondents referred to ES co-production through high human and social capitals (in average 3,62 and 3,69 out of the 1–5 scale), while physical and financial capitals had less relevance (in average 2,22 and 2,23). The significance of differences among answers was tested per stakeholders groups (SH.GROUPS) and per ES and Table 3.7 presents how *Social Capitals* involved in ES co-production depend on the ES considered.

Looking at the anthropogenic capitals involved in specific ES, some interesting differences can be observed in the graphs of Fig. 3.14. Considering the human capital, the highest values are expressed for C2 Environmental education (4,33) and C6 Esthetic beauty (4,17), while the lowest are related to P2 drinking water (2,5) and C8 artisan products (2,5). Social capitals are highest for C2 Environmental education (4,67) and C1 Eco-tourism (4,56) and lowest for P2 Drinking water (2,33) and C8 Artisan products (2,75). Physical capital: highest in P2 Drinking water (3,67) and C1 Eco-tourism (3,11); lowest are R2 Climate regulation (1,50) and C4 Intrinsic value (1,60). Financial capital: highest is P2 Drinking water (3,67) and C1 eco-tourism (3,00); lowest for C4 Intrinsic value (1,60) and C5 Sense of place (1,90).

Regarding ES Categories, Cultural services are mostly co-produced by Social (4,2) and Human capitals (3,9), while values of Physical and Financial capitals are lower (respectively 2 and 2,2). Regulating services account first Human (3,2) and then Social (3) capitals, while Physical and financial capitals both correspond to the value of 2. Provisioning services see a growing relevance of Physical and Financial capitals, with the first being the most relevant (together with financial capitals—3,2) and the second having a value of 3, together with human capitals (the lowest value among the categories).

Stakeholders' Dependency, Benefits, and Access to Decision-Making

The questionnaire (section D) addressed the topic of stakeholders' dependency for ES, the benefits they derive from the ES, and the access to decision-making regarding those ES. Figure 3.15 shows the normalized answers from the whole sample of respondents, looking at the five defined stakeholder groups.

The importance for the ES was discussed as the dependency of stakeholders on the existence of the specific ES they play a role in. Results in this section showed great relevance for *Production* for the P1 Agricultural products (90%), C8 Artisan

Table 3.7 Results of the Chi-square test on the distribution of answers on the capitals involved in ES co-production, per Stakeholders groups (SH.GROUPS) and per Ecosystem Service (ES)

Variables		df	p-value	Variables		df	p-value
SH.Groups	Human capital	16	0.6908	ES	Human capital	44	0.1627
SH.Groups	Social capital	16	0.2228	ES	Social capital	44	0.0327
SH.Groups	Phisical capital	16	0.0650	ES	Phisical capital	44	0.2164
SH.Groups	Financial capital	16	0.0798	ES	Financial capital	44	0.3291

ANTHROPOGENIC CAPITALS INVOLVED IN CO−PRODUCTION

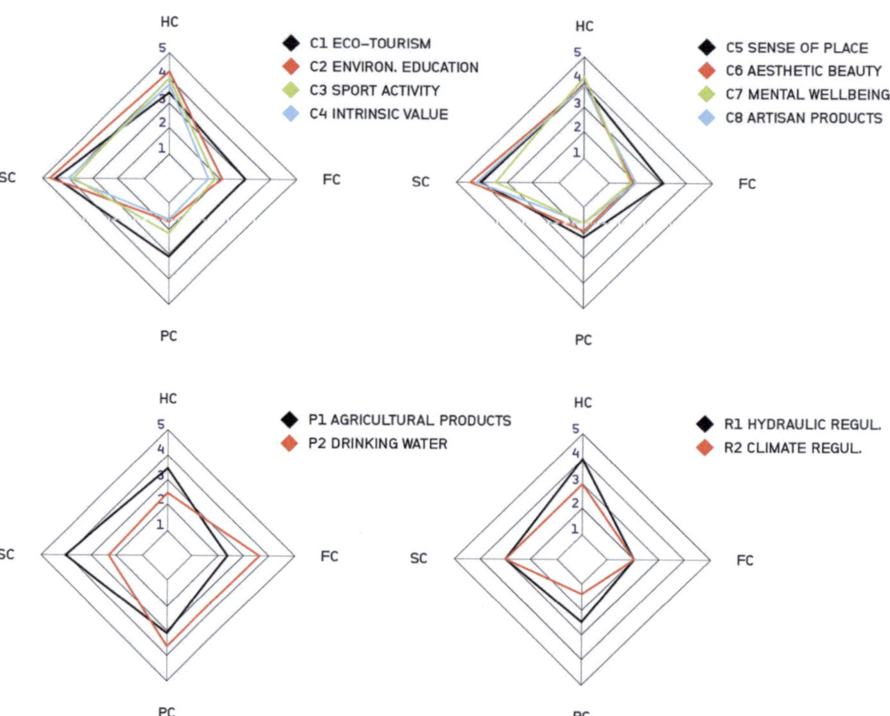

Fig. 3.14 Anthropogenic capitals involved in the co-production of Ecosystem Services as stated by local actors: Human capitals (HC), Social capitals (SC), Physical capitals (PC), Financial capitals (FC)

products (86%), and P2 Drinking water (75%). The Stakeholder group *School and Research* was considered dependent on C2 Environmental education (61%) and partly on C4 Intrinsic value (44%). *Tourism and commerce* had general low values in terms of dependencies, with 45% related to C1 Eco-tourism. *Planning and administration* have also generally low values with mid-rates regarding R2 Climate regulation (67%) and R1 Hydraulic regulation (50%). Finally, *Society* have mid-values for C5 Sense of place (55%), C4 Intrinsic value (50%), R2 Climate regulation (50%), C3 Sport activity (48%), C2 Environmental education (42%), and C6 Esthetic beauty (40%).

In terms of the benefits that stakeholder groups receive from ES, we observed relevant changes in respect to dependency. Stakeholders belonging to the *Production* group generally have lower benefits (36% of stakeholders are considered as dependent but only the 22% receive benefits from ES). The group *School and research* also showed a decrease of 14% to 25%. On the other hand, stakeholders within the *Tourism and commerce* group were selected as beneficiaries of benefits with a value almost twice as high as that of dependency (35% versus 19%). As for the two other

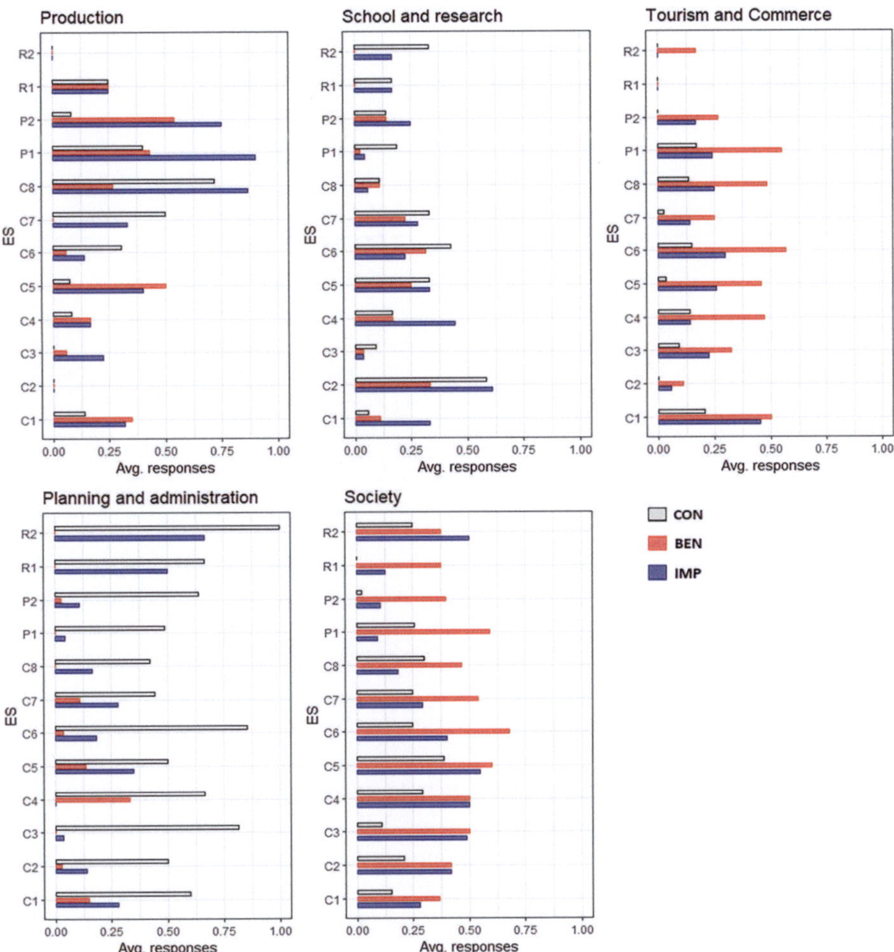

Fig. 3.15 Perception of questionnaire respondents on stakeholder groups regarding how much they consider the Ecosystem Service important for them (IMP), the benefits they get (BEN), and their contribution to decision-making (CON)

groups, *Planning and administration* decreased from 23% to 7%, while society increased from 33% to 48%.

Looking at the contribution to decision-making, results show also here a discrepancy in respect to dependency values. Table 3.8 presents the analysis looking at the difference between contribution to decision making (CON) and dependency on ES (IMP). In these terms, we can notice how the *Production* group does not have access to decision-making as it should, regarding dependence on ES, with an overall value of −15%, which increases to over 50% when considering specific provisioning services. The *School and research* group presents a balanced situation (−0,11%), while *Tourism and Commerce* and *Society* show a total negative value of, 11 and 12%,

Table 3.8 Differences in respondents' perception between Dependency on the Ecosystem Service and access to decision-making

ES	PRO	SCH	TOU	PLA	SOC	TOT
C1	−18,06%	−27,78%	−25,00%	32,41%	−12,50%	−12,35%
C2	0,00%	−2,78%	−5,56%	36,11%	−20,83%	−0,93%
C3	−22,22%	5,56%	−12,96%	77,78%	−37,50%	−1,23%
C4	−8,33%	−27,78%	0,00%	66,67%	−20,83%	0,93%
C5	−32,50%	0,00%	−22,50%	15,00%	−16,25%	−12,22%
C6	16,67%	20,37%	−14,81%	66,67%	−15,28%	8,02%
C7	16,67%	5,56%	−11,11%	16,67%	−4,17%	0,93%
C8	−15,00%	5,56%	−11,67%	25,56%	11,67%	2,22%
P1	−50,00%	14,44%	−6,67%	44,44%	16,67%	5,74%
P2	−66,67%	−11,11%	−16,67%	52,78%	−8,33%	−7,87%
R1	0,00%	0,00%	0,00%	16,67%	−12,50%	0,00%
R2	0,00%	16,67%	0,00%	33,33%	−25,00%	5,56%
	−14,95%	−0,11%	−10,58%	40,34%	−12,07%	−0,93%

respectively. On the other hand, the stakeholder group *Planning and administration* is seen as having most of the relevance in decision-making, while being poorly dependent on most of the ES.

Self-Perception in Decision-Making

After rating the contribution of other stakeholders to decision-making, the respondents to the questionnaire were asked to value their own influence as well. The data shown in Fig. 3.16 represent how respondents stated the contribution of other stakeholders in decision-making (A), how they rated their own contribution (B) and finally the rate of under−/over-estimation (C) obtained dividing the value of A by B and displayed by a Cartesian heat map.

It is noticeable that most stakeholder groups over-estimate their own role when answering about themselves compared to what others think about them. The *Production* group in particular showed an over-estimation with a rate of +0,2 in relation to C5 Sense of Place. *School and research* over-estimate their role for C4 Intrinsic value (+0,2), for C8 Artisan products (+0,3) and P1 Agricultural products (+0,2). *Tourism and commerce* showed +0,2 for C1 Eco-Tourism; 0 for C2 Environmental education; +0,1 for C5 Sense of place; +0,1 for C8 Artisan products and *Society* 0,2 for C3 Sport activity; 0,3 for P1 Agricultural products; 0,1 for P2 Drinking water; and 0 for R1 Hydraulic regulation.

On the other hand, the main under-estimation is related to *Planning and administration*, which under-estimate their role for R2 Climate regulation (−2,9), C6 Esthetic beauty (−2,1), C3 Sport activity (−2), and C4 Intrinsic value (−1,7). The group of *Production* also under-estimates its role in regards of C8 Artisan products (−1,3).

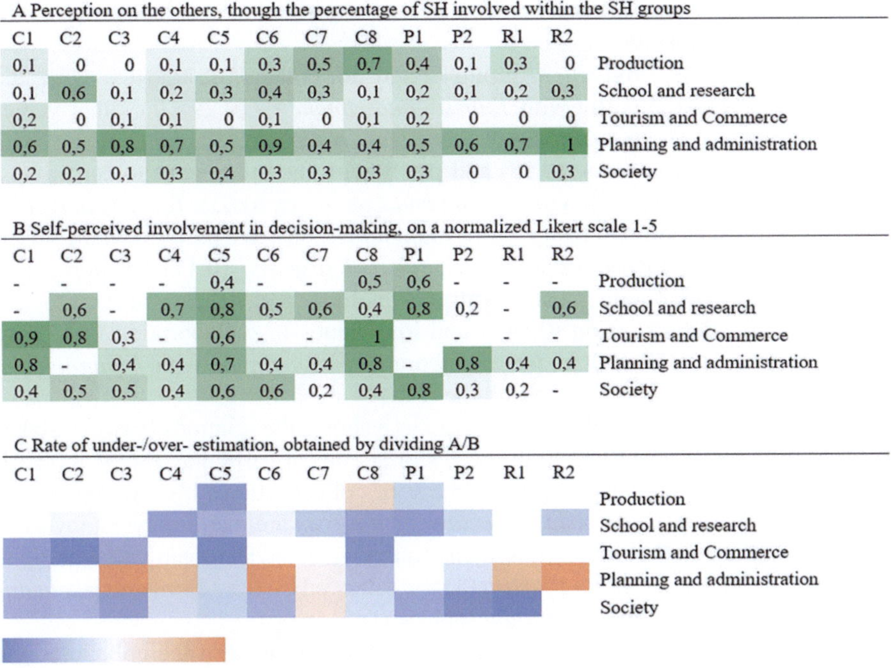

A Perception on the others, though the percentage of SH involved within the SH groups

C1	C2	C3	C4	C5	C6	C7	C8	P1	P2	R1	R2	
0,1	0	0	0,1	0,1	0,3	0,5	0,7	0,4	0,1	0,3	0	Production
0,1	0,6	0,1	0,2	0,3	0,4	0,3	0,1	0,2	0,1	0,2	0,3	School and research
0,2	0	0,1	0,1	0	0,1	0	0,1	0,2	0	0	0	Tourism and Commerce
0,6	0,5	0,8	0,7	0,5	0,9	0,4	0,4	0,5	0,6	0,7	1	Planning and administration
0,2	0,2	0,1	0,3	0,4	0,3	0,3	0,3	0,3	0	0	0,3	Society

B Self-perceived involvement in decision-making, on a normalized Likert scale 1-5

C1	C2	C3	C4	C5	C6	C7	C8	P1	P2	R1	R2	
-	-	-	-	0,4	-	-	0,5	0,6	-	-	-	Production
-	0,6	-	0,7	0,8	0,5	0,6	0,4	0,8	0,2	-	0,6	School and research
0,9	0,8	0,3	-	0,6	-	-	1	-	-	-	-	Tourism and Commerce
0,8	-	0,4	0,4	0,7	0,4	0,4	0,8	-	0,8	0,4	0,4	Planning and administration
0,4	0,5	0,5	0,4	0,6	0,6	0,2	0,4	0,8	0,3	0,2	-	Society

C Rate of under-/over- estimation, obtained by dividing A/B

C1	C2	C3	C4	C5	C6	C7	C8	P1	P2	R1	R2	
												Production
												School and research
												Tourism and Commerce
												Planning and administration
												Society

Over-					Under-estimation

Fig. 3.16 Cartesian heatmap representing the rate between the weight respondents gave to the category and the self-perception on contribution to decision-making. Data are presented only for Ecosystem Services which Stakeholders stated to have a role in. Over-estimation means that the Stakeholder group perceive higher role in decision-making than what stated by the rest of local actors

Collaboration among SH groups

Respondents:	Answ: Production	Answ: School and…	Answ: Tourism an…	Answ: Planning an…	Answ: Society
Production	0,43	0,09	0,29	0,05	0,11
School and research	0,40	0,47	0,03	0,53	0,30
Tourism and Commerce	0,63	0,17	0,25	0,29	0,34
Planning and administration	0,19	0,17	0,11	0,54	0,13
Society	0,17	0,11	0,06	0,33	0,17

Fig. 3.17 Collaboration among Stakeholders groups. The number express the average amount of Stakeholders (normalized value related to Stakeholders group) selected by respondents in the questionnaire

Collaboration Across Scales

Finally, the analysis turned to the dimension of collaboration, exploring how interactions among stakeholders occur across different scales of action. The questionnaire section E tackled the topic of collaboration among stakeholders and "Managers" in ES co-production, which defined networks of collaboration regarding the provision of specific ES. Fig. 3.17 shows relations among stakeholders organized by stakeholder groups.

Specifically, the *Production* group stated to collaborate mostly with the *Production* (0,43) and partly with *Tourism and commerce* (0,29), the other values are lower than 0,2. Respondents from *School and research* collaborate mostly with *Planning and administration* (0,53) and partly with the other stakeholder groups, while very low value is related to *Tourism and Commerce* (0,03). *Tourism and commerce* stated collaboration mostly with *Production* (0,63), and partly with the others categories, and the lowest value for *School and research* (0,17). *Planning and administration* collaborate mostly with stakeholders from the same category—*Planning and administration* (0,54)—and low values of collaboration (<0,2) are stated for the rest. Finally, *Society* collaborates partly (0,33) with *Planning and administration* and barely (<0,2) with the rest.

Looking at the single stakeholders, Fig. 3.18 shows the relations between stakeholders, distributed according to the scales of action. In particular, the municipalities are the major actor in play, with 13,8% of collaboration (42% as source and 58% as target). Also, *Bars, restaurants and agritourism* are major players, involved in 12,7% of the collaborations (mainly as source—73%). The main collaborations are related to the ES *C1—Eco-tourism* (31%), followed by *C2—Environmental education* (13%) and *C8 Artisanal products* (11%).

Regarding the spatial dimension, it can be seen that most of the collaborations are at interest to stakeholders at the local scale (83,8%), while regional and inter-regional stakeholders are involved in only 14,6% and 1,5% of the collaborations respectively. *Planning and administration* and *School and research* are the groups that mostly connect the local with the regional scale, including the 80% of stakeholders that are interested by beyond-local scale connections. The inter-regional-scale network regards only stakeholders from the *Tourism and commerce* group.

3.3.3 Lessons Learned: Role of Local Actors in Landscape Transformations

This study sought to include the perspective of social actors in landscape planning through the application of the ES co-production framework to a local case study. Our findings confirm that stakeholders associate great values to local ecosystems and identified several ES people derive from nature in the local case study. From the list of services defined by the local focus group, almost all of them were related to a co-production rate from respondents, through different roles: Society and the Education system were mostly stated as users, while production, tourism and commerce sector should be addressed as managers of the ES. As a general result of the analysis, the theoretical framework proved useful in highlighting social roles in landscape transformation, that, when properly integrated into planning, can support a sustainable and resilient development (Gret-Regamey, 2008).

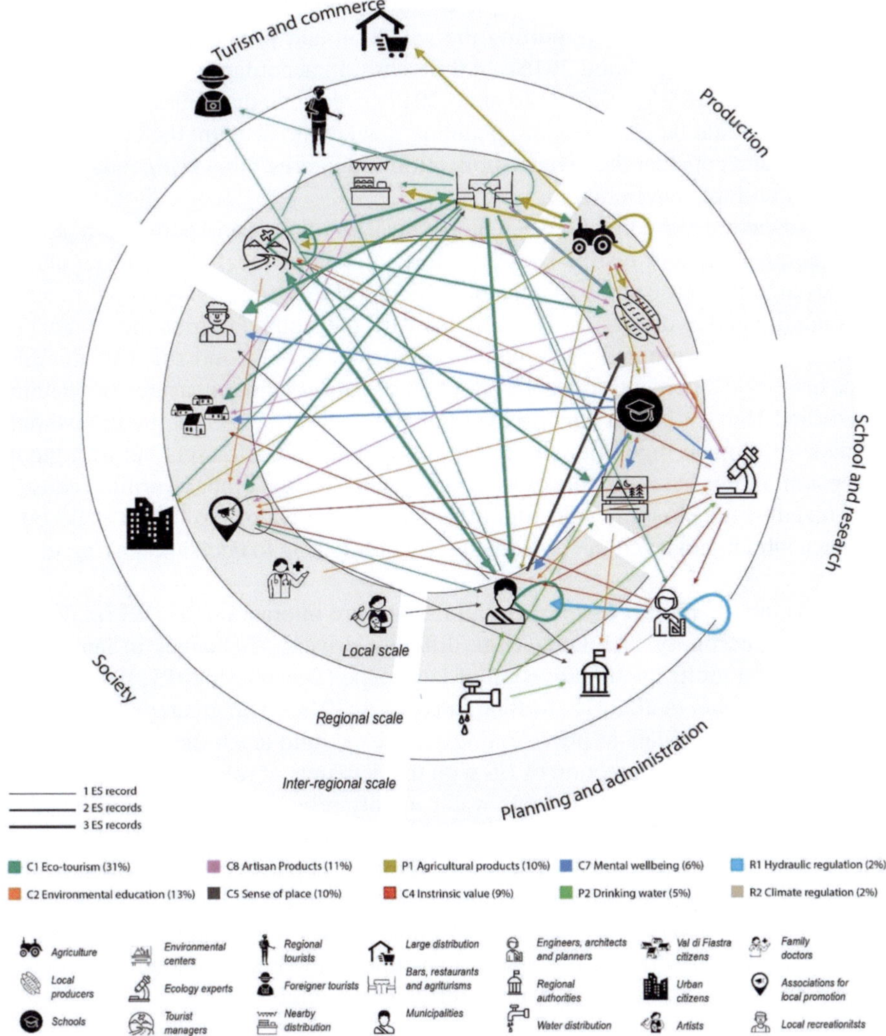

Fig. 3.18 Ecosystem Services stakeholders in the local case study. The stakeholders are organized in stakeholder groups and scale of action

As we know, local actors take roles in the co-production of ES through different kind of resources, knowledges, and expectations they invest in natural capitals (Palomo et al. 2016). Our empirical exercise demonstrated that human and social capitals have the greatest role in the co-production of services in rural environment. Among the others, social actors referred to education or knowledge (human capitals) as drivers for interaction with natural capitals, and to values and norms (social capitals) as incentives for their businesses. The exercise proved useful not only in showing how stakeholders act physically on the environment (physical co-production), but also how anthropic capitals shapes the perception of a service (cognitive co-production) supporting the valuation and protection of the environment (Fischer and Eastwood 2016). In this sense, in accordance with other studies on relational values (Arias-Arévalo et al. 2017), we argue that pluralistic valuation approaches should be included into planning to take into account the complexity of landscapes and consider the diverse motivation and interests that brings social actors to interact with the environment.

Looking at the close link between biophysical conditions and population perception, the questionnaire results showed a low interest of local actors in regulation services in a rural case study that is not particularly exposed to risks (both in terms of air quality and hydraulic aspects). In this way, the study confirms the connection between risk exposure and perception of regulatory service and suggests the relevance of people's perception for the adoption of conservation measures in planning (Lewis and Harvey 2001). Among the regulatory services, special mention should be made of climate regulation, a service not linked to local assets but to a global dimension. In this case, the role of local ecosystems for the global warming cause is recognized by the planning and administration stakeholder group (voted as ES preferences), which, however, under-estimate its contribution to decision making in this service.

Embracing the multiple aspects of human–nature interaction, the ES framework offers the opportunity to acknowledge different drivers of changes in landscapes and supports a multifunctional vision of landscape (Díaz et al. 2015; Felipe-Lucia et al. 2018; Fischer et al. 2015; Mascarenhas et al. 2014). The greater role of physical and financial capitals in provisioning services should teach the need to balance the enhancement and protection of ES with the demands of related anthropic activities. As the infrastructure of potable water mainly relies on physical and financial capitals, also regional agricultural systems require physical support for their resilience toward global challenges. The same argument applies to Eco-tourism, which, given the role played by physical and financial co-production, might be classified as provisioning service, in accordance with other studies (e.g., Pueyo-Ros 2018). In this sense, we argue that the framework can support integrated assessment in planning by defining specific support in the different planning sectors.

3.3.4 Cultural Values of Landscapes

The empirical exercise on the local case study underscored the usefulness of the ES co-production approach to display cultural values of rural landscapes. Those values are becoming more important as European policies (among others, *European Landscape convention, 2000)* call on governments to recognize landscapes as an essential component of people's lives and expression of their cultural and natural heritage. The inclination of local stakeholders for ES such as Esthetic beauty or Mental well-being shows the relevance of intangible benefits in the relation human–nature. Those benefits are often not caught by planning (de Groot et al. 2010a, b;

Ruckelshaus et al. 2015) but we argue that they should be pushed ahead as indispensable elements of landscape planning and could provide a rich basis for development strategies for inland areas.

This could lead to a realignment of the cultural heritage preservation agendas, especially in areas recently affected by catastrophic events where cultural elements are seen as the main connections for displaced populations to their original areas (EdT 2018). The prominent rate of co-production regarding sense of place as an ES found in our study shows awareness among local actors of the interdependence of nature and communities in sense of belonging to a place. Reconstruction strategies must therefore consider this interaction when planning emergency housing or the future development of rural settlements (EdT 2018; Sargolini et al. 2022).

With regard to tangible cultural services, strategies on inland areas often consider tourism as a mantra for the development of rural areas, as it can support economic diversification and has the potential to generate income and employment (Aretano et al. 2013; Petrosillo et al. 2007; Pueyo-Ros 2018). Nevertheless, the focus group did not express great preferences for Eco-Tourism, with an even lower value when considering the Civil Society focus group. On the other hand, in terms of co-production it is a quite engaging ES, being the most selected, with a high rate of managers (66%), not only from the tourism and commerce sector but also from planning and administration. For these reasons development strategies should support eco-tourism as a potential growing sector, while taking into account the roles of stakeholder's categories and the values they connect to it (Aretano et al. 2013).

3.3.5 Insights on Collaborative Planning and Access to Decision-Making

The study highlights several implications of ES co-production on relations among stakeholders. The three main findings regard: the discrepancy between dependency and decision-making on the ES; the divergency between self and other's perception on the service; the network of collaboration connecting social actors. Those relations are proved to affect the equity of the distribution of ES (Palomo et al. 2016) and need to be considered much stronger in planning as they influence how ecosystem structures eventually turn into benefits (Fischer and Eastwood 2016). The discrepancies between dependency and benefits for ES as well as between dependency and access to decision making need to be accounted as possible indicators of environmental injustices and lack of equity in distribution of benefits (Calderón-Argelich et al. 2021; Fischer et al. 2015).

Considering the stakeholders' dependency, benefits and access to decision-making, the investigation in the rural case study reveals how social actors related to Production sectors are the most dependent on ES. This is the case especially when focusing on tangible benefits, as the provision of water and agricultural products, or the cultural value of artisan products. They are the most dependent, while sharing the benefits with the Tourism and commerce sector, and the rest of society. This discrepancy on perspectives gets even more significant when analyzing the access to decision-making, where the role of stakeholders is not recognized in the management of the ecosystems, as their dependency for the service would suggest. We stress here that better access of the production sector to the management of the ecosystems might support a higher distribution of ecosystem benefits and give insight to planning in the direction of environmental justice (Calderón-Argelich et al. 2021).

The second main result considers the difference between self and other's perception of their own role in decision-making. This aspect is crucial in research on ES co-production, as the way people perceive themselves shapes the co-production process (Fischer and Eastwood 2016). In accordance with Jericó-Daminello et al. (2021), our findings confirm that the self-perceived role is frequently higher than the attributed role. As perception mismatches might lead to conflicts in stakeholders management (Zoderer et al. 2019), planning should support the raising of the awareness of territorial actors on the actual roles they have in the co-production of ES.

As for the last aspect considered, the study demonstrated the effectiveness of co-production in displaying the dense network of collaborations around the provision of ES. Two main outcomes can be synthetized from the results: the prominence of the local dimension as a scale of interaction, and the relevance of municipalities as actors connecting the diverse stakeholders. The former includes interactions within the production sector, collaborating mostly within itself and partly with tourist and commerce, and through the institutional actors, where school and research collaborate mostly with planning and administration (and planning and administration with itself). The latter has a downside related to the poor horizontal collaborations among actors, except a few with associations and local producers. In accordance to Opdam et al. (2015a), this demonstrates how using ES in planning has a strong multisectoral potential and stimulates the engagement of actors with diverging backgrounds. Nevertheless, collaboration and social learning can be fostered only through the support of novel assessment and design tools (Opdam et al. 2015a) and collaborative planning must be open to the integration of innovative tools for ensuring participation.

Box 3.4 Capturing Landscape Subjectivity Through Co-Production
Landscape is the creation of society within a specific territory (Sereni 1961), embodying the ambiguity between the "real country" and its representation. Planning often navigates the tension between objective ecological realities and the subjective perceptions of local actors.

While the spatial analysis focuses on biophysical indicators to understand the "objectivity of reality" in socio-ecological systems, this approach explores the "subjectivity of perception" using social science methods. The study introduces a framework for analyzing landscapes through ES co-production, applied in this book to the local case of Fiastra Valley in Le Marche Region.

Engaging local stakeholders via focus groups and interviews, the study gathers qualitative data on their perceptions and experiences. It evaluates the anthropologic role in ES co-production by examining stakeholders' roles and the various capitals (natural, social, human, financial) involved. Analysis highlights dependencies, benefits, access to decision-making, self-perception, and collaboration across scales.

The ES co-production framework effectively includes social actors, enhancing the relevance and inclusivity of landscape planning and management. Capturing subjective perceptions fosters greater stakeholder engagement and cooperation, crucial for sustainable development. The Fiastra Valley case study offers valuable insights that can inform similar methodologies in other regions.

By integrating objective biophysical data with subjective social insights, this approach provides a comprehensive view of landscapes. It highlights the potential of the ES co-production framework to enhance the role of social actors in landscape planning and management, offering a model for more effective and inclusive regional development.

3.3.6   Strengths and Shortcomings in Using the Concept of ES Co-Production in Analyzing Landscapes as Social-Ecological Systems

The integration of the ES co-production framework in landscape planning proved useful in highlighting the role of social actors in the management of ES at landscape scale. The concept is useful in connecting the ecological and social sphere by focusing on links between ES interdependencies and the perception of stakeholder involvement. In this way, integrating it into planning has the advantage of emphasizing relations among actors and accounting possible environmental inequities deriving from the provision of the services. Social evaluation methods are more and more implemented in planning, and participatory processes are explicitly addressed by policies as in the case of *The new Leipzig Charter, 2020.*[4] However, the implementation of social analysis requires large economic and human resources, which are not often available in contexts such as non-urban settings. Moreover, the distance of rural areas from places of decision making makes them often refractory and less willing to participate.

---

[4] The New Leipzig Charter provides a key policy framework document for sustainable urban development in Europe. The Charter promotes the concept of 'integrated development' policy and set out the key principles behind it. The document is available at:

https://ec.europa.eu/regional_policy/sources/docgener/brochure/new_leipzig_charter/new_leipzig_charter_en.pdf

The study also showed a great ability to capture cultural ES and assess the cultural values of landscapes. The inclusive methodology allowed the analysis to be directed toward the benefits that participants associate with ES, and this directly directed the work toward cultural ES. Nevertheless, cultural services are more related to people's perceptions, and it is not possible to exclude a bias of respondents in representing an entire category of stakeholders. This aspect requires further research. A combination of social network analysis (SNA) and descriptive statistics might support the selection of stakeholders (Jericó-Daminello et al. 2021) toward the definition of a more representative sample. However, participants' answers might not always represent the category but personal experience (e.g., out of the professional environment) and this weakness could be overcome expanding the number of participants.

Further research might enlarge the territorial scope of the analysis to urban case studies, as well as more peripheric areas, for example, mountain areas or national parks. This would allow the analysis of the differences in social roles in areas more related to the demand of ES (e.g., urban areas) or to the offer of ES (e.g., rural/peripheric areas) (Baró et al. 2017). In this way the study can offer a picture of territorial interdependencies, looking at power relations and spatial differences in access to decision-making (Gebre and Gebremedhin 2019). Furthermore, a focus group with regional authorities might account the discussion about trade-offs related to ES co-production and therefore support a more equal distribution of benefits.

3.4 The Landscape as a Social-Ecological Assembly

This concluding section frames the landscape as a social-ecological assembly, emphasizing its hybrid nature as both material and relational. In this perspective, territories emerge as dynamic configurations of ecological processes, social practices, and governance arrangements, offering a framework for interpreting the conclusions drawn from the two empirical studies developed in Le Marche Region. The first study adopts a spatially explicit approach based on ES mapping to highlight territorial interdependencies, spatial gradients, and functional configurations. The second study focuses on the perceptions, roles, and interactions of local stakeholders, applying the ES co-production framework to examine how people actively shape and are shaped by the ecosystems they inhabit.

Together, these two perspectives reveal the dual nature of landscape: as both a biophysical substrate structured by ecological and economic processes, and a lived space co-produced through social values, governance structures, and everyday practices. While Study 1 contributes to the understanding of regional patterns of supply, demand, and trade-offs, Study 2 foregrounds the importance of perception, equity, and collaboration in landscape planning. These conclusions support the broader aim of the book: to integrate ecological reasoning into territorial cohesion strategies and offer place-based tools for sustainable and inclusive regional development.

3.4.1   Towards a Recognition of Territorial Interdependencies

The ES assessment proved to be a powerful tool to analyze landscapes as social-ecological systems. The study highlighted bundles of ES supply and demand further explaining them as landscape units associated to local socio-economic assets. This allowed to propose management practices for sustainable landscape development. Furthermore, ES supply and demand were explored in terms of budgets within urban poles and inland areas as classified by the SNAI. The exercise underlined existing dependencies of urban toward inland systems in terms of environmental resources.

ES patterns are mostly shaped by spatial socio-cultural conditions, especially land use, demography, and income. Integrating these features in cohesion strategies can support regional governance in the path toward more equitable and sustainable development. Analyzing interdependencies beyond the urban–rural dichotomy enabled the study to recognize the polycentric characteristic of the regional case study which also presents urban conditions in mountain forest bundles. As for trade-offs between services, they were found especially in agricultural areas, between the supply of provisioning and regulating services, while synergies characterize the mountain systems where a set of regulating and cultural services are supplied. In relation to this and the global pressures facing inland ecosystems, we offered suggestions for sustainable landscape planning.

We demonstrated how areas of demand differ from areas of supply and concentrate in the most densely populated areas. We suggest that cohesion policies should embed a place-based approach integrating local characteristics in the strategies for regional development. This needs to be pursued by involving stakeholders, whose perspective is needed to ensure a balanced and sustainable development within regions.

This work contributed to the main objective of developing a mapping approach for landscapes as social-ecological systems, which is capable of highlighting spatial interdependencies between local systems. In particular, recommendations to planning can be offered in terms of (1) the integration of multisectoral governance; (2) the development of a territorial approach through systems of municipalities; (3) the contribution of an ecological perspective to territorial cohesion. Above all, this third aspect makes it possible to create a new narrative on spatial interdependencies, in which inland areas can be central to the provision of ES.

3.4.2   Accounting People's Perception in Landscape Planning

To our knowledge, Sect. 3.3 represents the first explicit application of ES co-production framework to landscape planning though an empirical work on a real case study. Results show that rural actors associate great cultural values to local landscapes, defining a series of ES they receive from ecosystems. Those services are actively co-produced by a different range of stakeholders both as managers, acting on the state of ecosystems, and users, benefiting from the services and valuing them. Moreover, using the ES co-production framework in landscape planning can highlight implications on stakeholder relations, stressing the extent actors depend on and benefit from the service. The comparison of those relations to the access to decision making can give an inside to the equity of the distribution of services among stakeholders and support environmental justice in regional planning.

The comparison between self-perception and perception by others regarding access to decision-making showed how there is a general over-estimation of own roles, with the exception of the planning and administration sectors, which should take better account of the relevance of their position in the management of ES. The integration of the ES co-production framework in planning proved useful in visualizing the collaboration networks among social actors, with municipalities being the main actor on the local scale.

Through the methods of social analysis, this work addressed the main objective of developing a framework for analyzing the role of local stakeholders through the concept of ES co-production. The integration of stakeholders in landscape planning allowed to highlight dependency and benefits relationships between society and nature, deriving implications for landscape planning on stakeholders' relationships and access to decision making. In addition to the relevant consequences in terms of environmental justice, the work proves the importance of a participatory approach to regional development, where local stakeholders are to be involved in the definition of strategies. This aspect will be developed specifically in the closing chapter of the book.

To conclude, this analysis, together with the biophysical analysis in Sect. 3.2, allow us to grasp the dual component of the landscape, on the one hand capturing the biophysical and economic territorial assets, and on the other investigating the perception of local actors in the co-production of these assets. The ES framework has proven to be a useful tool for capturing these aspects, exploiting the inherent and fundamental relationship between humans and ecosystems.

Chapter 4
Designing Territorial Futures

4.1 Introduction

This book offers a renewed perspective on cohesion policies by emphasizing the ecological value of local systems and their contribution to society (Magnaghi 2010). Through the lens of NCP, spatial interdependencies foster an ecologically oriented place-based approach where socio-economic patterns are coupled with environmental characteristics. The results thus argue for a transformative change not only placing at the center the areas of ES provision but also highlighting the role of anthropogenic systems co-producing those ES. Within this cause, this book advances NCP research in bridging the science-policy gap, also referred to as "implementation" gap (Levrel et al. 2017), addressing the unavailability of knowledge for practical implementation, as well as proposing an interdisciplinary framework for integrating different science perspectives and planning sectors. This book supports governance with inclusive and adaptive approaches to achieve sustainable cohesion policy at regional and local scale. On the one hand, the centrality of local systems, and on the other hand, the role of actors in the ES co-production allows for moving toward a participatory planning model based on the relationship between the social and the ecological spheres (Fig. 4.1).

As for the conceptual and methodological contribution, this book validates the potential of NCP in integrating different planning domains, facilitating discussions, and bridging between ecological and socio-economic aspects (Albert et al. 2014; Grêt-Regamey et al. 2017; Ruckelshaus et al. 2015). The research applies the NCP concept on the analysis of landscapes as social-ecological systems, in which the anthropogenic and natural components are interconnected to the point that they must be investigated within the same evaluation. The empirical study was successful in advancing the understanding of social-ecological systems though the concept of ES, while developing implications for landscape planning providing practical recommendation to achieve sustainability at a regional and global scale.

© The Author(s), under exclusive license to Springer Nature
Switzerland AG 2025
M. Giacomelli, *Ecologies of Cohesion*, SpringerBriefs in Geography,
https://doi.org/10.1007/978-3-032-01159-6_4

Fig. 4.1 Wolf on a prairie at a ski resort in the Sibillini Mountains, Le Marche, Italy. Image courtesy of Luca Castignani

This chapter synthesizes the outcomes of the research, organizing them around three main components. The first section concerns the scientific advances in methodology, mainly relating to the thriving research domain of ES in landscape planning. The second section provides suggestions for planning to capture the complexity of landscapes as social-ecological system. This served as a lens for addressing the core objective of the research, territorial cohesion, which is the topic of the third section. Here, insights are offered toward an ecological perspective on cohesion policies.

4.2 Conceptual and Methodological Contributions

The main innovation of this book lies in the dual analysis addressing the complexity of the landscape: on the one hand, the environmental assessment investigates the tangible components of socio-ecological systems, and on the other hand, the social analysis seeks to capture the perceptions of communities (Sargolini 2013). In this sense, the double dimension of the landscape is portrayed, and Sect. 3.2 investigates the objectivity of the biophysical dimension through spatial indicators, while Sect. 3.3 explores the subjectivity experienced by those who live the territory. The integration of two different working frameworks allows the construction of a complex methodology, which provides an important contribution to multidisciplinarity in landscape planning and governance.

The study framework builds on the integration of the multiple benefits ecosystems contribute to human well-being, including the social and cultural perspective on the ES cascade model (Haines-Young and Potschin-Young 2010; Martín-López et al. 2014). Yet, as benefits from nature do not occur independently but often require significant human contributions, the role of social systems is further analyzed though the ES co-production framework (Palomo et al. 2016). The landscape application is furthermore relevant because its scale is considered the most appropriate to the integration of social need and preferences to environmental assessments (Nogué and Sala 2018; Tallis et al. 2015). The contribution of this work to NCP science regards the use of novel methods for literature review as well as the integration of spatial and social evaluation methodologies.

In Sect. 3.2, the spatial dimension is assessed building on the concept of ES bundles (Raudsepp-Hearne et al. 2010), and developing it in the direction of the supply-demand perspective (Baró et al. 2017). The supply and demand definition (Villamagna et al. 2013) allows the operationalization of the framework on territorial interdependencies, through budgeting operation (Burkhard et al. 2012). The novelty of this work lies in the application of the analysis on the inland areas framework, which allowed the assessment of interdependencies between urban poles and inland areas beyond the urban-rural dichotomy (Lucatelli et al. 2019). Furthermore, the landscape characterization of the identified social-ecological systems represents a useful application of the ES bundles research on a Mediterranean region. The spatial assessment of supply and demand of a set of 12 locally relevant ES are a base for the development of a green infrastructure within the regional landscape planning context.

Conceptual and methodological contributions on the analysis of social systems within the landscape context are developed in Sect. 3.3. Here, the ES co-production concept is operationalized through focus groups and interviews on the value of nature for society in the local case study of Fiastra Valley. Building on the framework developed by Fischer and Eastwood (2016), as well as its implication on human's role in ecosystems and sustainability drawn by Palomo et al. (2016), the study uses the concept of co-production to read the connections between social and ecological components in the landscape as social-ecological system. In this frame, social relations are explored in order to understand power asymmetries and the distribution of ES benefits across the beneficiaries (Bennett et al. 2015). A recent application of the framework (Jericó-Daminello et al. 2021) highlighted discrepancies between how stakeholders perceive themselves as co-producers and how others perceive them. Focusing on stakeholders' dependency and benefits, the study offers a new tool for understanding stakeholders' relationships and access to decision-making, which can provide insights for more equitable and sustainable planning. Furthermore, the study demonstrated the effectiveness of co-production framework in displaying the dense network of collaborations related to ES dynamics.

4.3 Implications for Landscape Planning

Given the outlined methodological base, the research aimed at developing insights for regional governance and landscape planning to cope with drivers of transformation integrating social and ecological goals (de Groot et al. 2010a, b). The multidisciplinary approach described in the previous paragraph aimed at addressing the complexity of landscape as social-ecological systems. While landscape planning and management has historically often been based on a reductionistic understanding and linear models of the world (Holling and Meffe 1996), this research attempts to approach the complex interaction of different, and sometimes competing, land use characteristics in face of continuous biophysical as well as social changes, considering the local and the regional scale. Following the objectives stated previously, this book first accounted the knowledge domain, through the analysis of the state of the art on the integration of ES in planning (Sect. 1.4), and then developed a new functional landscape characterization through the concept of NCP (Sect. 3.2), including the role of social actors in the perception of landscapes through their role in co-production (Sect. 3.3). The following paragraphs present the main implication of the studies for landscape planning.

The literature review showed a thriving and growing research area, characterized especially by countries from the Global North, also addressing case studies in developing countries (Spangenberg et al. 2018; Teixeira et al. 2019). In the integration of ES in planning, a wide range of methodologies are adopted, ranging from spatial analysis to social data collection through questionnaires and interviews, or focus groups (Elbakidze et al. 2017; Giedych and Maksymiuk 2017). This variety of methods brings several application fields and tools. Among proactive planning, the most relevant concept results to be the one of the green infrastructure, applied both at city scale and regional scale (Arcidiacono et al. 2016; Artmann et al. 2017; Baró et al. 2017). Integration of ES in strategic planning includes strategies to protect or limit urban growth (Salata et al. 2020), while environmental and land use management show great emphasis on assessing the effects of land use changes. Research shows how analyses related to ES can support the planning process through new approaches to zoning with respect to multifunctionality and benefits offered by ecosystems (Geneletti 2013; Liu et al. 2019).

In Sect. 3.2, the application of ES bundles to map landscapes as social-ecological systems highlighted avenues of sustainable development for local landscapes in a Mediterranean regional case study. The study highlighted bundles of ES supply and demand further explaining them as landscape units associated to local socio-economic assets. The regional landscapes system was interpreted along a coastal-mountain gradient drawn by the rising altitude and decreasing population density (Queiroz et al. 2015; Quintas-Soriano et al. 2019). Those functional landscape units are mostly shaped by spatial socio-cultural conditions, especially land use, demography, and income. Analyzing interdependencies beyond the urban-rural dichotomy enabled the study to recognize the polycentric characteristic of the regional case study which also presents urban conditions in mountain forest bundles. Particularly interesting for regional decision-making, ES tradeoffs were found in agricultural areas, between the supply of provisioning and regulating services, while synergies characterize the mountain systems where a set of regulating and cultural services are supplied (Balzan et al. 2020; Felipe-Lucia et al. 2020). In relation to the global pressures faced by inland ecosystems, recommendations were offered for sustainable landscape planning:(1) preservation of the identity of inland systems through the support of local ecosystem management and responsible ecotourism strategies, (2) the enhancement of sustainable agriculture through small-scale farming and sustainable practices (also in relation to CAP), (3) pay special attention on multifunctional landscape management practices, in the local case study represented among the others by pastoralism.

The use of the ES co-production framework to visualize the role of local stakeholders within social-ecological systems (Sect. 3.3) allowed the integration of the social perspective in landscape planning, in respect to their dependency and benefits from ES as well as their access to decision-making. To our knowledge, this real case application represented the first explicit integration of ES co-production framework to landscape planning. Results show that rural actors associate great cultural values to local landscapes, visualized though a set of ES they receive from ecosystems (Díaz et al. 2018). Those services are both physically co-produced by actions on the state of ecosystems, as well as valued through the perception as users, benefiting from the services and valuing them (Palomo et al. 2016). The study thus proved the efficacy of NCP framework in visualizing how landscapes are co-produced by social and natural components, assessing both the physical social-ecological patterns and the perception of local communities (Sargolini and Gambino 2016). Incorporating the central role of people in landscape assessment, the study suggests that collaboration and social learning should be fostered through collaborative planning, integrating innovative tools for ensuring participation (Opdam et al. 2015).

4.4 Towards an Ecological Perspective on Cohesion Policies

While advancing environmental sciences on the application of the NCP concepts in landscape planning, this book aimed at providing regional governances with a scientific "environmental argument" in support of territorial cohesion. As described in the introduction, the process of urban growth led to increasing inequalities between regions (Rockström et al. 2009) and inland areas are undergoing a process of economic decline and depopulation (ESPON 2021). The Territorial Agenda 2030[1] encourages spatial planning and policymaking to promote a balanced and

---

[1] The Territorial Agenda 2030 underlines the importance of and provides orientation for strategic spatial planning and calls for strengthening the territorial dimension of sector policies at all governance levels. The document is based on the meeting of Ministers for Spatial Planning and Territorial Development and/or Territorial Cohesion 1 December 2020, Germany.
https://ec.europa.eu/regional_policy/sources/docgener/brochure/territorial_agenda_2030_en.pdf

harmonious territorial development between and within regions. To approach this challenge, this research read territorial interdependencies through the lens of NCP, emphasizing the role of inland systems in providing benefits to society.

The argument of territorial interdependencies arises from the awareness that the growth of urban settlements is based on the exploitation of natural capitals mostly coming from inland areas (Folke et al. 1997; Gebre and Gebremedhin 2019). Section 3.2 investigates the topic through the spatial analysis of ES demand and supply in Le Marche regional case study, characterized by growing polarization both in demographic and economic terms. As other Mediterranean areas, the region can be read through local systems of municipalities which are deeply different in terms of socio-economic profiles and spatial characteristics, ranging from remote mountain areas to urbanized valleys and coastlines (Calafati and Mazzoni 2009). The results highlighted a strong dependency of urban areas on inland systems concerning nearly the total set of 12 ES taken into consideration. The analysis further demonstrates how areas of demand differ from areas of supply and concentrate in the most densely populated areas. On the other side, systems providing services are located in areas remote from cities, where the great availability of natural resources support the delivery of ES. In this sense, the research produced a new environmental-based argument toward territorial cohesion where inland areas should no longer be defined by their condition of marginality and backwardness, but acquires a central position in the provision of benefits for the whole society. While cohesion policies often addressed disadvantaged areas though welfarist approaches based on "compassionate compensations" (Barca 2018), this book shows how inland systems should be supported for their role in the ES provision and the protection of biodiversity.

Nevertheless, the high aging indexes as well as the income gap toward the urban areas, draw urgent challenges for the resilience of inland systems. Actions in the direction of environmental justice should go beyond the redistribution of income or benefits but also include the role of local actors in decision-making as well as their cultural identities, values and knowledge systems (Pascual et al. 2014). The importance of local involvement is further underlined in the investigation of social systems though NCP co-production (Sect. 3.3). Here the results show discrepancies between dependency and decision-making, stressing the need for processes of social inclusion in landscape management. The exercise offers insights in terms of the equity of the distribution of ES (Palomo et al. 2016) and how ecosystem structures eventually turn into benefits (Fischer and Eastwood 2016). Stakeholders' perspectives are crucial to ensure a balanced and sustainable development within regions and its integration in planning constitute civic practices that can directly generate societal benefits (Krasny et al. 2014).

When including the perspective of local actors, the notion of participation can be helpful. Difficult to define (Pellizzoni and Osti 2003), two factors primarily structure participatory processes: the will to act and the extent of the possibilities of intervention. Referring to the sphere of policy generation processes, the extent of citizens' possibilities of intervention in decision-making processes can be structured according to the "scale of participation" proposed by Arnstein (1969). This scale shows different rungs of intervention ranging from *Nonparticipation* to different degrees of *citizen power*. Empowerment is indeed crucial in cohesion policies, in the sense of increasing citizens' capacities of elaboration and intervention as well as defining their own development horizons (Bobbio and Pomatto 2007). The ability to increase citizens' active participation also means inclusion of those actors generally excluded from decision-making processes. By assigning roles to stakeholders of ES co-production, this research goes in the direction of enabling local actors to assume responsibility and expand the possibility of influencing decision-making.

Through the integration of people's perspective, the book proposes new tools for the design of place-based policies, also referring to the process of "policy territorialization." According to Galli et al. (2013), the territorialization of policies has a double characteristic: first, it means moving from a sectoral to a territorial and integrated logic, linking not only the different sectors but also individuals and operators, second, it means strengthening bottom-up activities and collaboration. With this aim, this work follows the objective of overcoming the cognitive crisis of the central state caused by the regionalization of the contemporary countryside, fostering the active and conscious participation of the different actors in the rural areas (Pellizzoni and Osti 2003). Participation and integration are, therefore, key words for a new model of rural development policies as well as for the application of a multifunctional planning perspective.

In many ways, these concepts are well known and debated within the cohesion policy discourse, although they are far from being applied in the concrete application of policies (Barca et al. 2012). Overcoming this gap, this research builds on participation and integration to create an ecologically oriented place-based approach, combining local social-ecological assets through the lens of landscape. This approach can overcome the idea of fit-for-all solutions designed by central governments, which characterized past welfare policies for the inland areas in Italy (Viesti 2016), and instead support the development of approaches based on local landscape characteristics. In accordance with The Territorial Agenda 2030, this place-based approach builds on local capabilities and aim to promote innovative ideas through the interaction of local knowledge with external actors, in line with the existing territorial interdependencies.

To conclude, suggestions toward new ecological cohesion policies can be drawn from the conducted analysis as well as the theoretical conceptualization behind this research. Beyond the crucial base of information given by spatial analysis, involvement and participation become necessary in the case of local development and integrated projects. The limits of a top-down, technocratic approach have been made visible by 20 years of European cohesion policy applications and a new orientation toward the construction of active communities is necessary. It becomes evident therefore how policies should support processes rather than project, supporting the empowerment of local actors for the self-development of their own paths.

Appendices

1.1 Appendix A: Individual Ecosystem Services

1.1.1 Provisioning Services

The ES P1 Cereal production was mapped according to actual production (supply) and actual consumption (demand). *The supply map* highlights a main production strip in the mid-low hilly area from south to north. No supply is recorded in the south-western mountain areas while lower values concern the rest of the inland areas. The service supply also decreases on the coast due to a major level of urbanization. *The demand map* highlights a high consumption associated with the coastal municipalities and to the urbanized valleys penetrating the inner region.

As for the Cereal production, the ES P2 Wine production was mapped following indicators of actual production (supply) and actual consumption (demand). *The supply map* reports concentration of productions in areas recognized as DOC and DOP, such as Rosso Conero (in the area of Ancona), Rosso Piceno (in the low Ascoli Piceno Province), and Verdicchio (area of Jesi and Matelica). The rest of the Municipalities present a diffuse a moderate supply level, apart from mountainous areas, which are not prone to the cultivation of wine. The *demand map* is related to population density and accounts higher values in the coast and in the main urban poles.

The ES P3 Pastoral production was mapped through the indicator of cheese production (supply) and consumption (demand). *The supply map* shows that productions are mostly located in the mountain part in the southwest and in the north of the region, while low productions are mapped along the coast. Pastoral productions are thus distinctive among local productions, characterizing themselves as high hill and mountain activities. Similarly to ES P1 and P2, the *demand map* is related to population density and reports higher values in the coast and in the main urban poles (Fig. A.1).

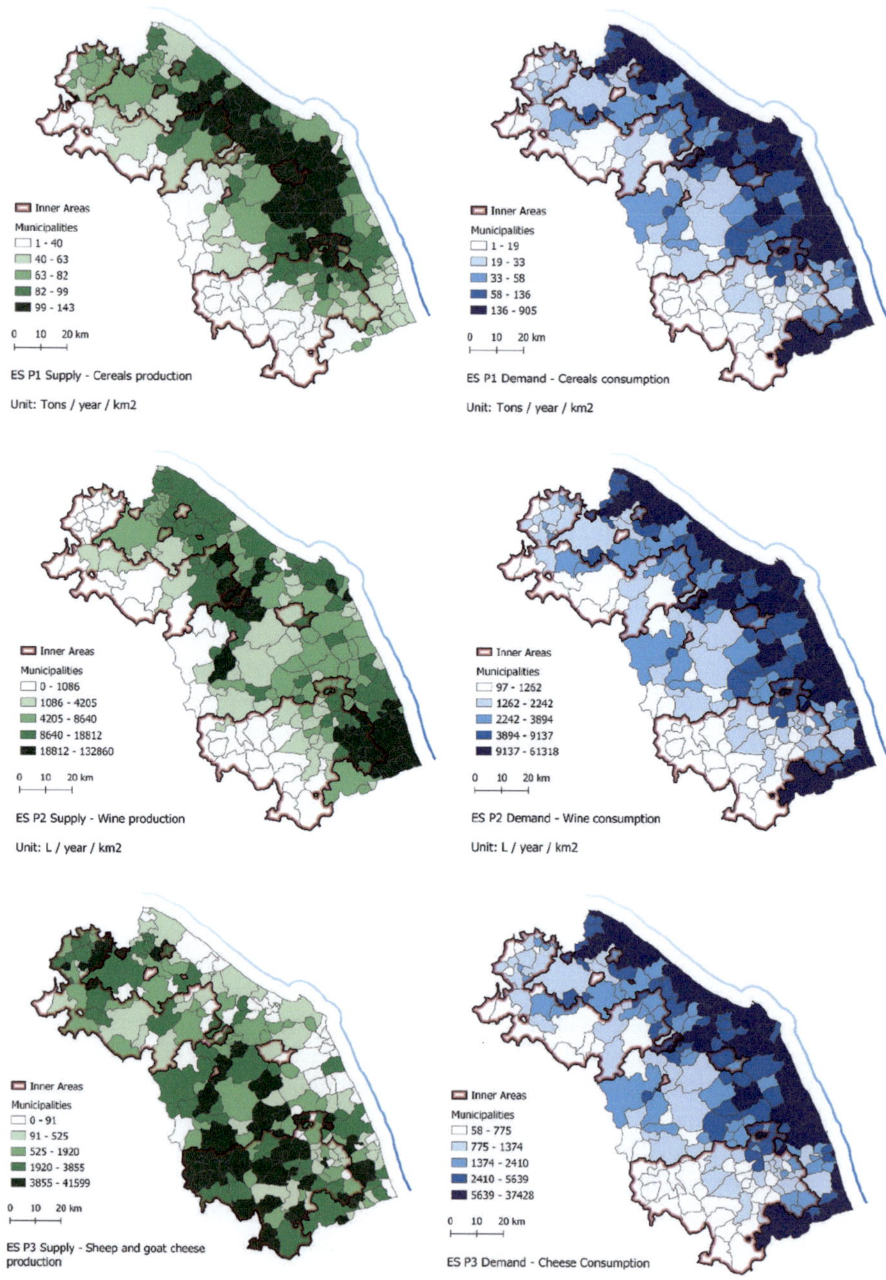

Fig. A.1 Spatial patterns of ES P1 Cereal production (up), P2 Wine products (center), and ES P3 Pastoral products (down), indicators. The map on the left indicates ES supply and on the right indicates ES Demand

The ES P4 Drinking water relates to the water distributed by the local aqueducts. The supply indicator is defined as the concessions allowed by Le Marche Region for the capture of water for drinking purposes while the demand refers to the actual water delivered by the network to the population. The *supply map* highlights different profiles for the north and south of the region: the north offers a rather uniformed distribution with hotspots municipalities corresponded to the main water withdrawal points, while the south shows a strong supply from the inland areas. The *demand map* relates to population density with higher values on the coast and main urban centers.

The ES P5 Hydro power was mapped through the nominal power of regional hydroelectric plants (supply) and the energy consumption by population and companies (demand). The supply map highlights hotspots of production in the mountain area in the southwest of the region and other municipalities along the valley. As in the case of P5 drinking water, the supply is distributed throughout the municipalities without major unbalances. The *demand maps* draw the actual consumption along the areas of higher population densities, on the coast and in the main urban poles (Fig. A.2).

1.1.2 Regulating Services

ES R1 Hydraulic regulation relates to the flood risk from extreme events (demand) and the role of vegetation in rainfall infiltration (supply). The *supply map* shows high values for the mountain forested areas while low values are mapped in low hilly fields. The *demand map* highlights three spots of higher pressure, respectively, in a southern, central, and northern hilly areas.

The ES R2 Soil protection refers to the loss of soil due to water erosion and the role of ecosystems in retaining nutrients. The *supply map* shows how highest soil retention values correspond to areas covered by woodland and pastures. The demand map indicates a hotspot in the southwest, with general higher values on the hilly belt (Fig. A.3).

The map of R3 Crop Pollination refers to the service provided by a wide range of insect species that with the help of wind allow plants and trees to develop fruits, vegetables, and seeds. The *supply map* refers to the relative pollination potential and is related to the presence of forests and trees in the inner belt of the region. The *demand map* reveals the dependency of cultures from pollination especially in the hilly area devoted to fruit trees and partially oilseeds.

R4 Climate change regulation concerns processes related to atmospheric chemical composition, particularly to the greenhouse effect. The *supply map* related to the absorption of CO_2 by ecosystems and, as for the ES crop Pollination, it shows higher values in the forested inner belt of the region. The *demand* is related to the emission of CO_2 and the map gives higher values for the municipalities of the coast, together with the first hilly belt, hosting most of the regional industrial activities (Fig. A.4).

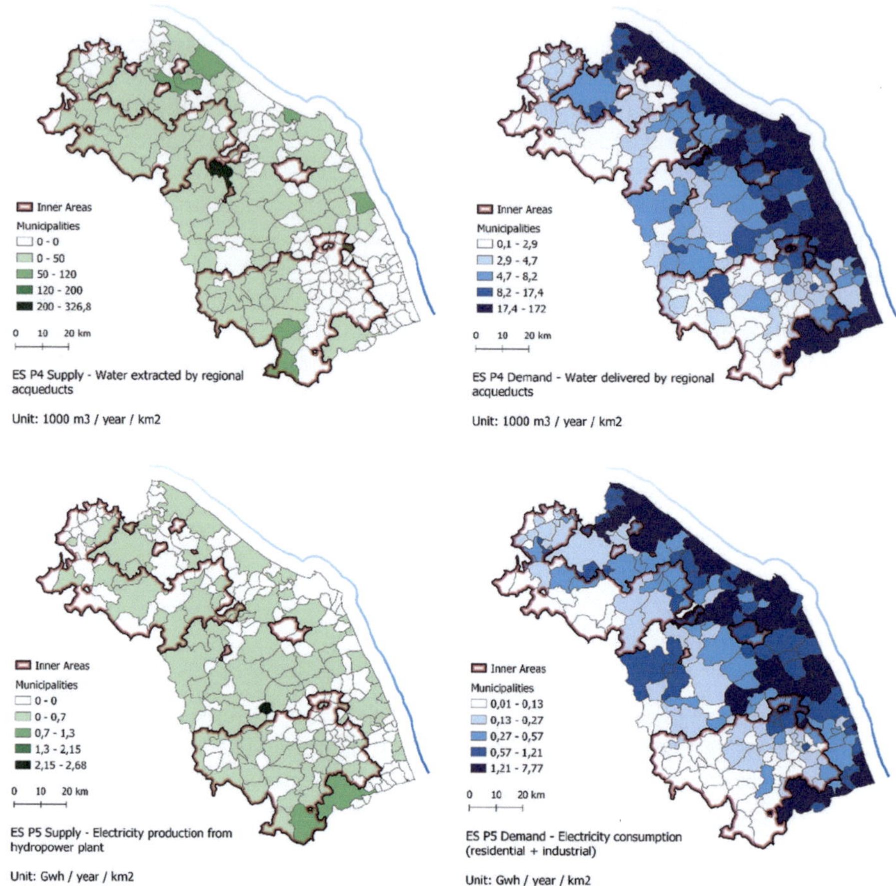

Fig. A.2 Spatial patterns of ES P4 Drinking water (up) and ES P5 Hydro power (down) indicators. Left: ES supply, right: ES Demand

1.1.2.1 Cultural Services

The ES C1 Eco-tourism relates to the recreational pleasure people derived from natural or cultivated ecosystems. The *supply map* relates to the indicator of available hiking paths and shows higher values in mountain municipalities and protected areas. *The demand* is related to visitors in eco-touristic structures and interests mainly the coastal area with some inland hotspots.

As well as C1, the activity related to C3 Mushroom picking reflects a leisure interest related to ecosystems. In this case the demand refers not to incoming tourists but to local residents. The *supply map* is related to geographical condition allowing habitats for mushrooms and has higher values in forested mountain areas

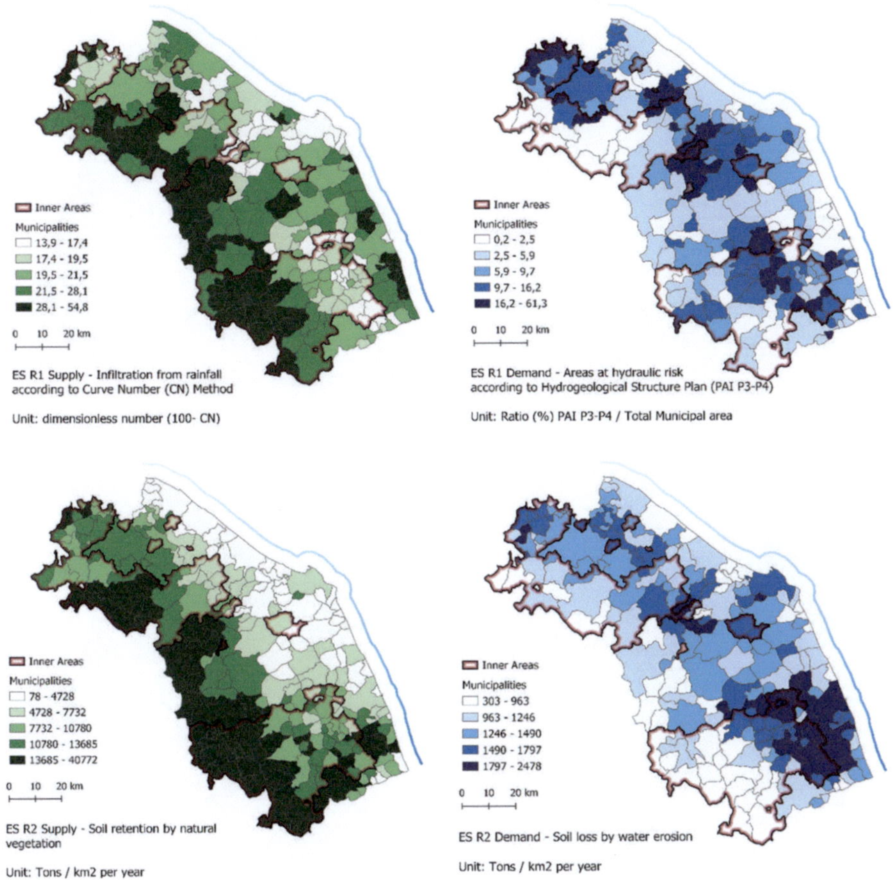

Fig. A.3 Spatial patterns of ES R1 Hydraulic regulation (up) and ES R2 Soil Protection (down) indicators. Left: ES supply, right: ES Demand

coinciding with the inner belt. The *demand* is connected to the distribution of mushroom license in the regional territory and gives higher values in territorial hotspots spread throughout the region (Figs. A.5 and A.6).

C2 Environmental education relate to the process of learning from nature both in academic programs and in informal settings. For this reason, the *supply map* shows an equal distribution of spots connected to the Regional Education Centers and didactic farms, without any clear difference among inland areas and more urban contexts. The *demand* is centered in more populated areas, especially the coastal poles hosting younger population.

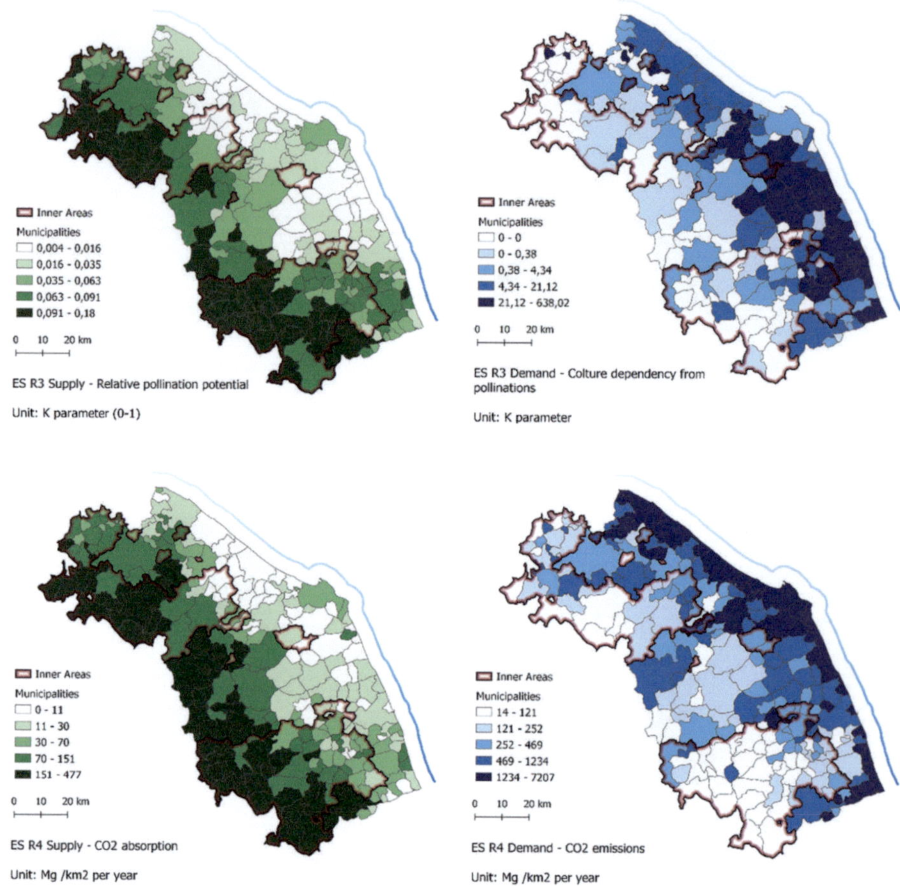

Fig. A.4 Spatial patterns of ES R3 Crop Pollination (up) and ES P4 Soil Erosion regulation (down) indicators. Left: ES supply, right: ES Demand

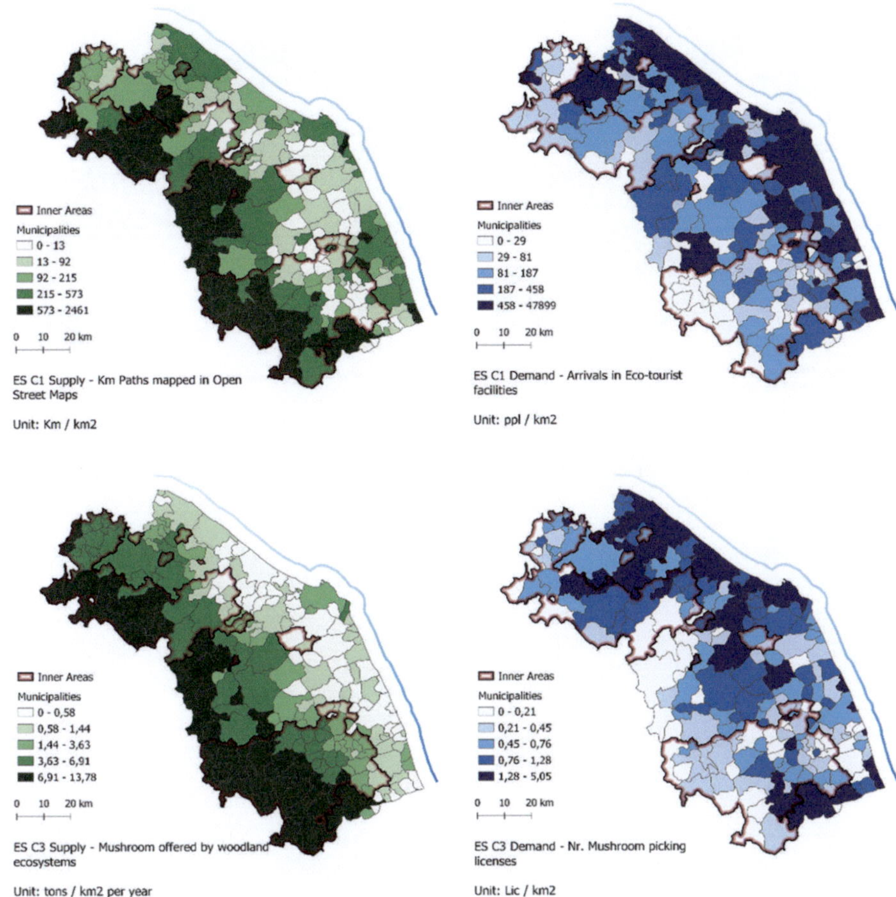

Fig. A.5 Spatial patterns of ES C1 Eco-Tourism (up) and C3 Mushroom picking (down) indicators. Left: ES supply, right: ES Demand

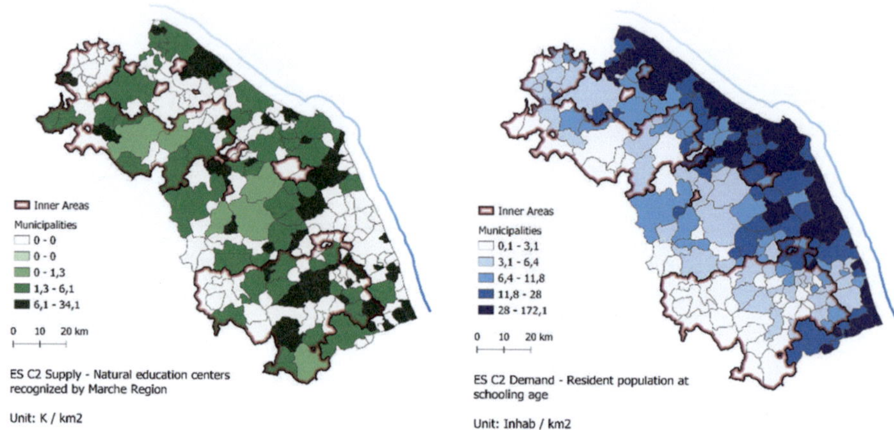

Fig. A.6 Spatial patterns of C2 Environmental Education (down) indicators. Left: ES supply, right: ES Demand

1.2   Appendix B: Individual Socio-Economic Indicators

The results of the socio-economic characterization are shown in respect to the three categories of indicators. The *social indicators* include S1 Demographic index, S2 Social and material vulnerability index, and S3 Income. The *economic indicators*, refers to the three economic sectors E1 Primary sector, E2 Secondary sector, and E3 Tertiary sector, while the *land use indicators* include L1 Artificial surfaces; L2 Agricultural surfaces; L3 Forests and seminatural areas.

1.2.2.1   Patterns of Social Indicators

S1 Demographic index relates to the aging of the population and shows clear unbalances between inland areas and urban poles. S2 Social and material vulnerability, accounting seven different dimensions of social and material vulnerability, gives a patchy picture of the region, with high values of the indicator mostly in the belt areas. The last social indicator, S3 Per capita taxable income, highlight again a homogeneous correspondence between the income of population and the inland areas classification, with coastal and urban municipalities having the highest income and the inland areas, the lowest values (Fig. B.7).

1.2.2.2   Patterns of Economic Indicators

The maps show the distribution of the economic indicators, respectively, illustrating employment in the primary sector (E1), secondary sector (E2), and the tertiary sector (E3). The first indicator shows a significance prominence of agriculture, forestry, and fishing sector for the south of the region. The indicator related to manufacturing activities indicates high value for the municipalities close to the coast. Finally, the accommodation and recreational activities highlight peak values in the coastal southern municipalities (probably connected to seaside tourism) and individual hotspots in the inner side of the region (Fig. B.8).

1.2.2.3   Patterns of Land Use Indicators

The selected indicators focus on three specific land uses: L1 artificial surfaces; L2 agricultural surfaces; L3 forests and seminatural areas. The L1 shows the incidence of artificial surfaces on the total municipal value and highlights a clear difference between the coastal municipalities and the in-land territories, with valley strips from the coast toward the mountain in the center of the region. L2 highlights a central hilly area dedicated to agriculture, with very low value in the mountain municipalities. L3 shows a mountain belt of woodland and seminatural areas, with lower values in the area of Monte Conero and the southern province (Fig. B.9).

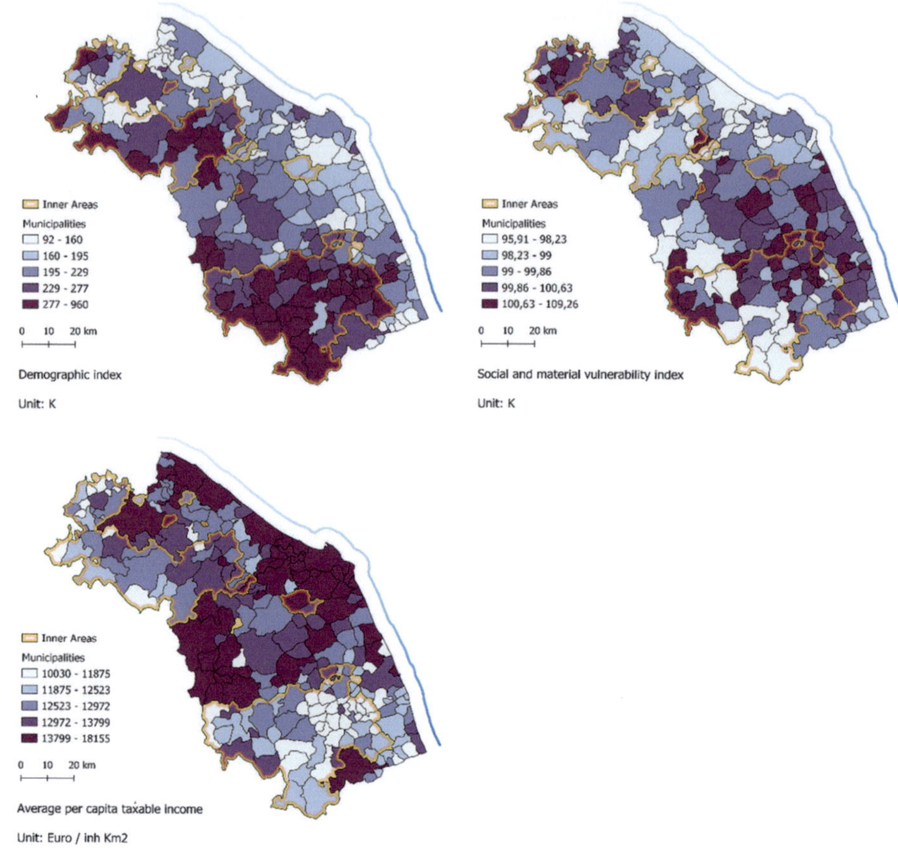

Fig. B.7 Spatial pattern of the socio-economic indicators

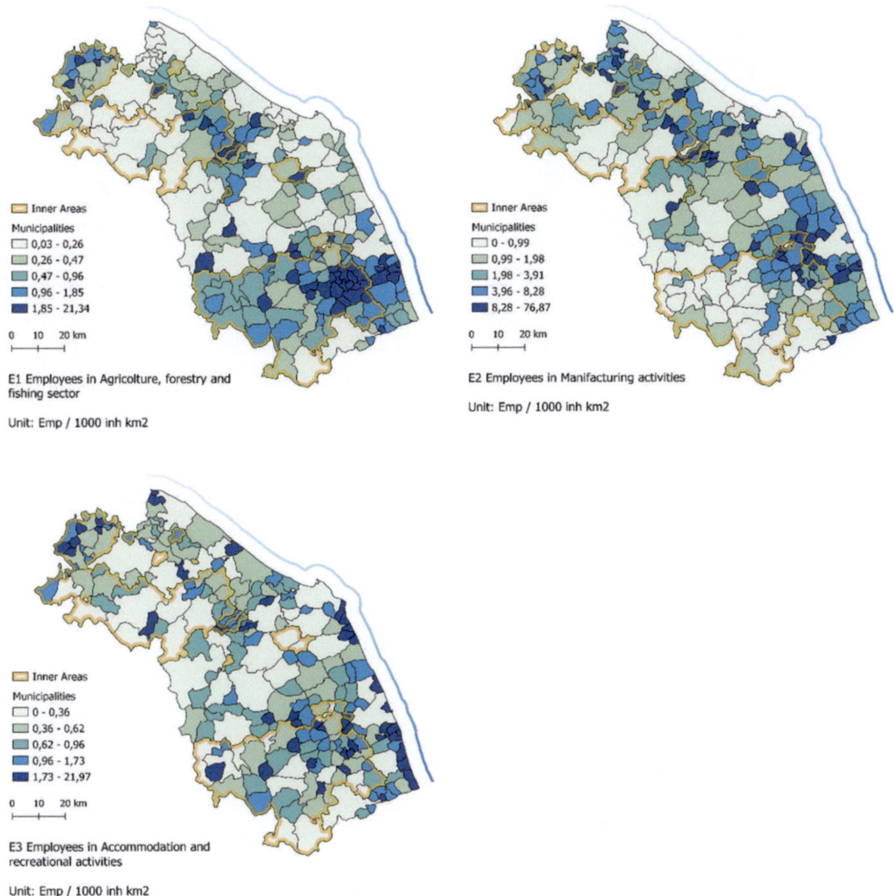

Fig. B.8 Spatial patterns of economic sectors (E1 primary, E2 secondary, and E3 tertiary)

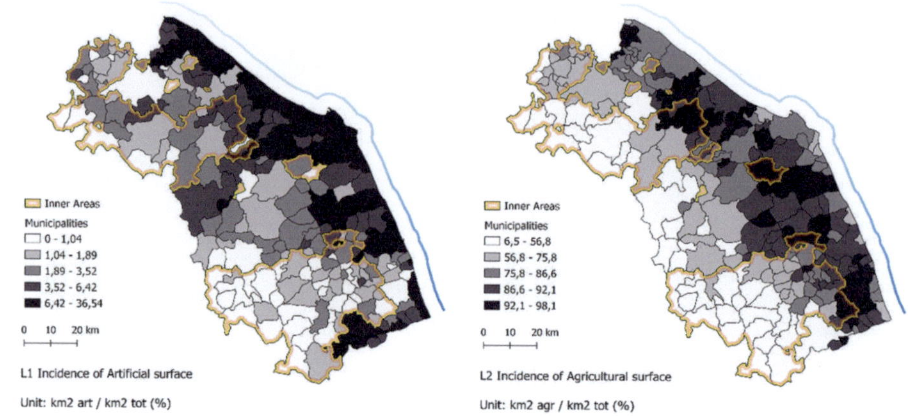

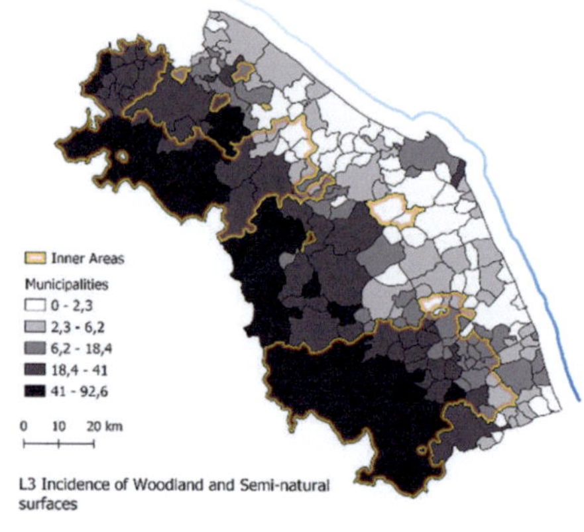

Fig. B.9 Spatial patterns of the Land Use indicators

1.3   Appendix C: R Code Study 1

```r
R.input <- read.csv2 ("C:/Users/…")


# I devide the ES dataframe from the Socio-economic dataframe
R.input <- data.frame(R.input)
ES_Data <- select (R.input, P1_S, P1_D, P2_S, P2_D, P3_S, P3_D,
P4_S,    P4_D, P5_S, P5_D, R1_S, R1_D, R2_S, R2_D, R3_S, R3_D,
R4_S, R4_D, C1_S, C1_D, C2_S, C2_D, C3_S, C3_D)
SOCIO_Data <- select (R.input, SNAI, S1, S2, S3, E1, E2, E3,
L1, L2, L3)
ES_Data_S <- select (R.input, P1_S,    P2_S, P3_S, P4_S,    P5_S,
R1_S, R2_S,    R3_S, R4_S, C1_S, C2_S, C3_S)
ES_Data_D <- select (R.input, P1_D, P2_D, P3_D, P4_D, P5_D,
R1_D, R2_D, R3_D, R4_D, C1_D, C2_D, C3_D)

library(tidyverse)    library(dplyr) library(tidyr)
library(ggplot2) library(reshape2) library(ggthemes)
library(fmsb) library(gridExtra) library(grid) library(lattice)
library(scales) library(stats) library(corrplot) library("xlsx")
library("factoextra") library(vegan)

# I keep the input data in a separate dataset
ES_Data1 <- ES_Data
#Determine number of clusters
wss <- (nrow(ES_Data)-1)*sum(apply(ES_Data,2,var))
for (i in 2:15) wss[i] <- sum(kmeans(ES_Data,
centers=i)$withinss)
plot(1:15, wss, type="b", xlab="Number of Clusters",
     ylab="Within groups sum of squares")
# K-Means Clusterization
set.seed(123)
fit <- kmeans(ES_Data, 5) # 5 cluster solution
# get cluster means
aggregate(ES_Data,by=list(fit$cluster),FUN=mean)
# append cluster assignment
ES_Data1 <- data.frame(ES_Data, fit$cluster)
ES_Data1 <- rename (ES_Data1, CLUSTER = fit.cluster)
```

```r
#cluster number for each municipality
fit$cluster
head(fit$cluster, 4) #i look at the first 4
# Cluster size
fit$size
# Cluster means
fit$centers
fviz_cluster(fit, ES_Data, geom="point")+
  scale_colour_manual(values = c("violet", "darkgreen", "orange",
"blue", "green")) +
  scale_fill_manual(values = c("violet", "darkgreen", "orange",
"blue", "green")) +
  ggtitle("PRINCIPAL COMPONENT ANALYSIS")+
  annotate("text", x=7, y=-3, label= "C1 URBAN COASTAL")+
  annotate("text", x=2.5, y=0, label= "C2 CROPLAND")+
  annotate("text", x=0, y=3, label= "C3 CROPLAND AT
HYDRAULIC RISK")+
  annotate("text", x=-1, y=0, label= "C4 MOSAIC CROPLAND
FOREST")+
  annotate("text", x=-4, y=-3, label= "C5 MOUNTAIN FOREST")+
  theme_bw()


# Operation should be run only once
# N. 4 URBAN COASTAL (1)
# N. 3 CROPLAND (2)
# N. 1 CROPLAND AT HYDRAULIC RISK (3)
# N. 5 MOSAIC CROPLAND FOREST (4)
# N. 2 MOUNTAIN FOREST (5)
ES_Data1$CLUSTER <- ifelse(ES_Data1$CLUSTER==1, 3,
               ifelse(ES_Data1$CLUSTER==2, 5,
                      ifelse(ES_Data1$CLUSTER==3, 2,
                             ifelse(ES_Data1$CLUSTER==4, 1,
                                    ifelse(ES_
Data1$CLUSTER==5, 4,

                                          NA  ))))) # all other
values map to NA
SOCIO_Data$CLUSTER <- (ES_Data1$CLUSTER)


# related to clusters
CLU1 <- dplyr::filter(R.input,BUNDLE=="1")
CLU2 <- dplyr::filter(R.input,BUNDLE=="2")
CLU3 <- dplyr::filter(R.input,BUNDLE=="3")
CLU4 <- dplyr::filter(R.input,BUNDLE=="4")
CLU5 <- dplyr::filter(R.input,BUNDLE=="5")
clusters <- data.frame(colMeans(CLU1), colMeans(CLU2),
                       colMeans(CLU3), colMeans(CLU4),
                       colMeans(CLU5))
```

```r
# related to supply-demand (related to analysis with 11 ES)
CLU1_S <- select(CLU1,
"P1_S","P2_S","P3_S","P4_S","P5_S","R1_S","R2_S","R3_S","R4_S","
C1_S","C2_S",    "C3_S")
CLU1_D <-
select(CLU1,"P1_D","P2_D","P3_D","P4_D","P5_D","R1_D","R2_D","R3_
D","R4_D","C1_D", "C2_D","C3_D")
CLU1_SOCIO <- select (CLU1, "S1", "S2", "S3", "E1", "E2", "E3",
"L1", "L2", "L3")


# Same operation for CLU2, CLU3, CLU4, CLU5

CLU5_SOCIO <- select (CLU5, "S1", "S2", "S3", "E1", "E2", "E3",
"L1", "L2", "L3")
clusters_s <- data.frame(colMeans(CLU1_S), colMeans(CLU2_S),
                      colMeans(CLU3_S), colMeans(CLU4_S),
                      colMeans(CLU5_S))
clusters_d <- data.frame(colMeans(CLU1_D), colMeans(CLU2_D),
                        colMeans(CLU3_D), colMeans(CLU4_D),
                        colMeans(CLU5_D))
clusters_socio <- data.frame(colMeans(CLU1_SOCIO),
colMeans(CLU2_SOCIO),
                            colMeans(CLU3_SOCIO),
colMeans(CLU4_SOCIO),
                            colMeans(CLU5_SOCIO))

# If the p-value is < 0,05, then the correlation between x and y
is significant.
corr_S <-cor(ES_Data_S)
CS <- corrplot(corr_S, method = 'square', type = 'lower', diag
= FALSE)
corr_D <-cor(ES_Data_D)
CD <- corrplot(corr_D, method = 'square', type = 'lower', diag
= FALSE)
grid.arrange(CS, CS, nrow=1, top = textGrob("Correlation in ES
Demand and Supply",gp=gpar(fontsize=20,font=3)))

# If the p-value is < 0,05, then the correlation between x and y
is significant.
# I consider socio economic data
corr_SOCIO <- cor(SOCIO_Data)
corrplot(corr_SOCIO, method = 'shade', order="AOE")
```

```r
corr_TOT <- cor(R.input_to.plot)
corrplot(corr_TOT, method = 'color', type="lower")
#SOCIAL INDICATORS
# Correlation between budnles and S1 Demography (Result:
6.673e-13 -> very significant)
cor.test(SOCIO_Data$CLUSTER, SOCIO_Data$S1, method = "pearson")
# Correlation between budnles and S2 Vulnerability (Result:
0.2789 -> not significant)
cor.test(SOCIO_Data$CLUSTER, SOCIO_Data$S2, method = "pearson")
# Correlation between budnles and S3 Income (Result: 1.014e-06 ->
very significant)
cor.test(SOCIO_Data$CLUSTER, SOCIO_Data$S3, method = "pearson")
#ECONOMIC INDICATORS
# Correlation between budnles and E1 Agricolture (Result: 0.5408 ->
not significant)
cor.test(SOCIO_Data$CLUSTER, SOCIO_Data$E1, method = "pearson")
# Correlation between budnles and E2 Manifacturing (Result:
0.02881 -> significant)
cor.test(SOCIO_Data$CLUSTER, SOCIO_Data$E2, method = "pearson")
# Correlation between budnles and E3 Accommodation and recreation
(Result: 0.0004779 -> significant)
cor.test(SOCIO_Data$CLUSTER, SOCIO_Data$E3, method = "pearson")
#LAND-USE INDICATORS
# Correlation between budnles and L1 Artificial areas (Result:
2.2e-16 -> very significant)
cor.test(SOCIO_Data$CLUSTER, SOCIO_Data$L1, method = "pearson")
# Correlation between budnles and L2 Agricolture areas (Result:
2.2e-16 -> very significant)
cor.test(SOCIO_Data$CLUSTER, SOCIO_Data$L2, method = "pearson")
# Correlation between budnles and L3 Woodland areas (Result:
2.2e-16 -> very significant)
cor.test(SOCIO_Data$CLUSTER, SOCIO_Data$L3, method = "pearson")

# Correlation between bundles and SNAI classes (Result: 6.108e-11 ->
very significant)
# If the p-value is < 0,05, then the correlation between x and y
is significant.
cor.test(SOCIO_Data$CLUSTER, SOCIO_Data$SNAI, method = "pearson")

#I calculate diversity index (TOT)
DIV <- diversity(ES_Data, index = "simpson", MARGIN = 1, base
= exp(1))
#I calculate diversity index for SUPPLY
DIV <- diversity(ES_Data_S, index = "simpson", MARGIN = 1, base
= exp(1))
ES_Data1$Diversity <- DIV
```

```r
clusters_socio$Socio <- c("S1 Demography", "S2 Vulnerability",
"S3 Income",
                        "E1 Agricolture", "E2 Manifacturing", "E3
Accommodation and recreation",
                        "L1 Artificial areas", "L2 Agricultural
areas", "L3 Woodland and semi-natural areas")
  clusters_socio %>%
    ggplot()+
    geom_col(aes(x=colMeans.CLU1_SOCIO.,y=Socio),
            width = 0.12,
            alpha=0.5,
            position = position_nudge(y = 0.30),
            fill = "blue",
            colour ="black")+
    geom_col(aes(x=colMeans.CLU2_SOCIO.,y=Socio),
            width = 0.12,
            alpha=0.5,
            position = position_nudge(y = 0.15),
            fill="orange",
            col="black")+
    geom_col(aes(x=colMeans.CLU3_SOCIO.,y=Socio),
            width = 0.12,
            alpha=0.5,
            position = position_nudge(y = 0),
            fill="purple",
            col="black")+
    geom_col(aes(x=colMeans.CLU4_SOCIO.,y=Socio),
            width = 0.12,
            alpha=0.5,
            position = position_nudge(y = -0.15),
            fill="yellow",
            col="black")+
    geom_col(aes(x=colMeans.CLU5_SOCIO.,y=Socio),
            width = 0.12,
            alpha=0.5,
            position = position_nudge(y = -0.30),
            fill="darkgreen",
            col="black")+
    labs(title = "BUNDLES SOCIO-ECONOMIC CHARACTERISTICS")+
    theme_bw()
```

```r
clusters_s$Supply <- c("P1 Cereal production", "P2 Wine
products", "P3 Sheep products",
                       "P4 Drinking water", "P5 Hydro power", "R1
Hydraulic regulation",
                       "R2 Soil protection", "R3 Climate change
regulation",
                       "R4 Crop pollination", "C1 Eco-Tourism",
"C2 Environmental education",
                       "C3 Mushrooming")
s1 <-
  ggplot(clusters_s, aes(x=seq(1,360,by=360/
nrow(clusters_s)),y=colMeans.CLU1_S.)) +
  geom_bar(width=360/nrow(clusters_s),stat='identity',position =
"stack",colour=("grey90"),aes(fill=Supply)) +
  geom_hline(yintercept = 1 ,linetype="dotted")+
  geom_vline(xintercept = seq(37,396,by=360/
nrow(clusters_s)),linetype="dotted")+
  coord_polar()+
  guides(fill="none")+
  theme_void()

# Same code for plotting s2, s3, s4, s5

clusters_d$Demand <- c("P1 Cereal production", "P2 Wine
products", "P3 Sheep products",
                       "P4 Drinking water", "P5 Hydro power", "R1
Hydraulic regulation",
                       "R2 Soil protection", "R3 Climate change
regulation",
                       "R4 Crop pollination", "C1 Eco-Tourism",
"C2 Environmental education",
                       "C3 Mushrooming")
d1 <-
  ggplot(clusters_d, aes(x=seq(1,360,by=360/
nrow(clusters_d)),y=colMeans.CLU1_D.)) +
  geom_bar(width=360/nrow(clusters_d),stat='identity',position =
"stack",colour=("grey90"),aes(fill=Demand)) +
  geom_hline(yintercept = 1 ,linetype="dotted")+
  geom_vline(xintercept = seq(37,396,by=360/
nrow(clusters_d)),linetype="dotted")+
  coord_polar()+
  guides(fill="none")+
  theme_void()
```

```r
# Same code for plotting d2, d3, d4, d5

# 4.1.3) I print supply-demand graphs for each cluster
c1 <-   grid.arrange(d1, s1, nrow=1, top = textGrob("B1 Urban
coastal",gp=gpar(fontsize=15,font=1)))
c2 <-   grid.arrange(d2, s2, nrow=1, top = textGrob("B2 Cropland",
gp=gpar(fontsize=15,font=1)))
c3 <-   grid.arrange(d3, s3, nrow=1, top = textGrob("B3 Cropland
at hydraulic risk",gp=gpar(fontsize=15,font=1)))
c4 <-   grid.arrange(d4, s4, nrow=1, top = textGrob("B4 Mosaic
cropland forest",gp=gpar(fontsize=15,font=1)))
c5 <-   grid.arrange(d5, s5, nrow=1, top = textGrob("B5 Mountain
forests",gp=gpar(fontsize=15,font=1)))
#c6 <-   grid.arrange(d6, s6, nrow=1, top = textGrob("ES Bundle
6",gp=gpar(fontsize=15,font=1)))

# Print all together
c_tot <- grid.arrange(c1, c2, c3, c4, c5)
#ggsave(c_tot, "C:/Users/toma/documents/plot.png")
#LOCATION.PNG <- "C:/Users/toma/documents/ES-Bundles.png"
#png(c_tot, filename = LOCATION.PNG, width = 480, height = 480,
units = "px", pointsize = 12)

LOCATION <- "C:/Users/toma/documents/R-output.xlsx"
LOCATION.CSV <- "C:/Users/toma/documents/R-output-to_gis.csv"
write.xlsx(as.data.frame(ES_Data1), LOCATION, sheetName =
"ES-Mapping")
write.xlsx(as.data.frame(clusters), LOCATION, sheetName =
"clusters")
write.xlsx(as.data.frame(DIV), LOCATION, sheetName = "diversity")
write.csv(ES_Data1, LOCATION.CSV)

write.xlsx(as.data.frame(CLU1), LOCATION, sheetName ="Bundle 1")
write.xlsx(as.data.frame(CLU2), LOCATION, sheetName ="Bundle 2",
append=TRUE)
write.xlsx(as.data.frame(CLU3), LOCATION, sheetName ="Bundle 3",
append=TRUE)
write.xlsx(as.data.frame(CLU4), LOCATION, sheetName ="Bundle 4",
append=TRUE)
write.xlsx(as.data.frame(CLU5), LOCATION, sheetName ="Bundle 5",
append=TRUE)
```

1.4   Appendix D: Focus Group Materials

During the focus group, participants were asked to freely list benefits society derive from ecosystems, answering thought post-its to the following questions: (i) what does nature in the Fiastra Valley mean to you? what is the importance of nature for this area? (ii) what are the benefits that Fiastra Valley provides for your well-being/ to fulfill your organizational goals? (iii) what goods and products are given to society by nature in Fiastra Valley? (iv) how does nature sustain the local economy? The following tables present the content of the post-its, which were then translated into ES categories by moderators.

1.4.2.1   Group 1: Sunday 11 July 2021

ES Category	Post-IT Content
Environmental education	1. I benefici offerti dalla natura della Val di Fiastra sono molteplici: Ispirazione per la pratica didattica, motivazione, armonia ed. educazione alla bellezza 2. Una vallata "sociale" ha benefici psicologici e indubbiamente educativi 3. La Val di Fiastra è ambiente privilegiato di apprendimento 4. È bellezza che educa
Scientific (*together with environmental education)	1. Attraverso spazi di crescita, studio, osservazione
Eco-tourism	1. Creare un turismo alternativo 2. La natura offre opportunità di lavoro in ambito turistico alla riscoperta di radici profonde 3. Impatto sul turismo 4. Turismo esperienziale 5. B n b 6. La natura della Val di Fiastra sostiene l'economia attraverso il turismo (da potenziare) 7. Turismo 8. Ospitalità interattiva 9. Turismo 10. Ambiente naturale gradevole e non disagevole [accessibilità]
Sport activity	1. Piste ciclabili e pedonali nella vallata 2. Bike/e-bike

ES Category	Post-IT Content
Mental Well-being	1. La natura offre un benessere generale che rigenera "il corpo e lo spirito" 2. La natura nella Val di Fiastra sostiene l'economia attraverso il benessere di chi vi abita 3. Mi permette di avere ritmi più a misura d'uomo 4. Ippocrate diceva "fa che il cibo sia la tua medicina e la medicina il tuo cibo." Partendo dalla prima osservazione, possiamo concepire quanto la natura e i suoi frutti siano indispensabili per la salute di chi vive in un determinato ecosistema 5. Possibilità di pensare e di perdersi 6. Scoperta, cultura e benessere 7. Salute, ispirazione, equilibrio 8. Mi ritempra, riempie il mio animo di bellezza e di vita 9. Gli attimi di riflessione guardando la natura ricaricando le energie 10. Benessere, poesia, tranquillità 11. Favorisce il benessere 12. La Valle del fiastra offre un benessere generale 13. Ritmi umani 14. Natura = rinascita 15. Serenità, tranquillità
Intrinsic value	1. La natura è assolutamente intrinseca alla vita nella Val di Fiastra, parlare della sua importanza sembra quasi inadeguato perché non è qualcosa di altro rispetto al vivere in questi territori ma da forma e condiziona e permette o meno esperienze quotidiane 2. All'interno di una visione dicotomica uomo-natura, quest'ultima indubbiamente significa molto. Io credo però che dovremmo superare questa distinzione e considerarci come parte stessa della natura, che è un elemento imprescindibile per la nostra sopravvivenza. In quest'ultima prospettiva la natura non ha molto significato, è semplicemente tutto 3. Possiamo capire come la natura sostenga l'economia, soprattutto nella nostra vallata, attraverso un esercizio molto utile ma altrettanto inquietante. Immaginiamo una realtà distopica in qui tutti quegli elementi della natura che diamo per scontati (e infiniti) scompaiono. Forse così riusciremmo a percepire il suo valore, è essenziale. Inutile elencare i servizi che la natura offre al turismo, all'agricoltura, all'arte nella vallata, sarebbe una lista infinita
Biodiversity (*together with intrinsic value)	1. L'importanza della natura nella val di fiastra sta nel polmone verde e nella biodiversità che questo ecosistema offre 2. Spazio, diversità, ricchezza

(continued)

ES Category	Post-IT Content
Sense of place	1. L'impatto dell'ambiente sull'uomo, animali, piante e viceversa 2. La natura è l'aspetto principale che rende la vallata unica (colline, montagne, boschi, borghi, fiumi che combinano colori sapori odori ad ogni stagione) 3. Natura significa ambiente agricolo + tradizioni agricole, radici della nostra storia. Caratteristiche queste peculiari e fondamentali per il futuro 4. La natura nella val di fiastra è bacino di accoglienza e di cura della comunità che la compongono e la popolano. La natura è intreccio e trama delle nostre espressioni 5. Senso di appartenenza 6. Identità-relazione 7. Natura è territorio e tutto cioè che circonda i borghi abitati che li avvolge creando un'atmosfera unica di cui apprezziamo il valore solo quanto per diversi motivi legati al comportamento umano e la intacchiamo 8. Il bene più importante è la molteplicità e la ricchezza in anni di cura da parte Dei contadini sentinelle del territorio. Un patrimonio che va scomparendo 9. Identità in cui tutti ci riconosciamo 10. La natura è l'essenza del territorio della val di fiastra, il patrimonio comune
Sense of community (*together with sense of place)	1. Società = coesione 2. Aggregazione 3. Cittadinanza attiva 4. La natura, oltre che per la mera sopravvivenza e benessere di chi vive in un'area, è importante anche per definire un'identità, una cultura e un'immaginazione di una comunità abitativa 5. Nella val di fiastra la natura è prospettiva per il futuro
Aesthetic beauty	1. La bellezza innanzitutto 2. La bellezza 3. Nella val di fiastra la natura è educazione alla bellezza, all'armonia 4. Cosa significa la natura = bellezza, libertà, autonomia, armonia.. da gustare assaporare fodere e da educare
Source of inspiration (*together with aesthetic beauty)	1. È fonte di ispirazione 2. Panorami che arrivano all'anima 3. Contemplazione e silenzio che favoriscono l'autenticità e la ricerca del senso della vita
Artisan products	1. Possibilità di prodotti artigianali locali 2. Forni a legna con uso di pasta madre 3. Botteghe con produzioni artigianali locali 4. Punti vendita di prodotti artigianali locali 5. Tipicità enogastronomiche locali

ES Category	Post-IT Content
Agricultural products	1. Impatto sull'agricoltura e allevamento con le sue risorse 2. La natura della val di fiastra sostiene l'economia attraverso l'agricoltura (da rendere sempre più ecosostenibile) 3. Agricoltura 4. Attraverso una produzione agricola più attenta all'ambiente 5. Ambiente ricco di opportunità nell'attività agricola 6. Essendo un territorio di cultura agricola, credo che sostenga molto l'economia locale. Occorre una nuova visione per lasciare che ci sostenga sempre di più 7. Attraverso risorse alimentari
Air purification	1. Scarso inquinamento 2. Acqua aria suolo poco inquinati
R2 climate regulation	1. Attraverso risorse climatiche
Disservices (not included in the study)	1. Dopo il sisma la natura è più maligna e non sempre sostiene l'economia locale
Others not included within ES categories	1. Beni e servizi alla società forniti dalla natura 2. La natura fornisce i Beni di sostentamento (acqua, aria pulita, cibo Sano…) alla società 3. La natura è un capitale, e i "frutti" sono gli interessi. Se consumiamo il capitale lo depauperiamo, non avremo il sostegno del capitale iniziale 4. Offre Beni primari di qualità ed. opportunità in tutti i settori, basta crederci 5. Territorio della valle, ricco di natura e di storia, rimasto sufficientemente intatto nei secoli sicuramente di più rispetto ad altre aree del maceratese 6. Volano economico e di sviluppo ecosostenibile 7. Offre opportunità di lavoro (anche non troppe) ma in un ambiente lavorativo speciale 8. La possibilità di stare nella natura "facilmente" con tutte le conseguenze positive che la natura ha sull'individuo 9. Con l'aiuto di una adeguata pista ciclabile, si potrebbe: Godere meglio la natura, utilizzare il percorso per spostamenti dal capoluogo ai luoghi di lavoro

1.4.2.2   Group 2: Saturday 17 July 2021

ES Category	Post-IT Content
Eco-tourism	1. Credo che la natura in questo territorio per la sua società ed. economia locale rivesta un ruolo fondamentale, basti pensare ad esso come una delle risorse principali per l'economia locale attrazione per il turismo locale e non 2. Natura come meta di turismo per la sua bellezza 3. Turismo e attività 4. Esperienza di contatto con il paesaggio / paesaggio culturale (parco archeologico / parco Abbadia di Fiastra) (**) 5. Servizi di ristorazione con viste mozzafiato, 6. Viabilità, ristoro, infrastruttura

(continued)

ES Category	Post-IT Content
Sport activity	1. Servizi culturali legati agli spazi aperti naturali (es. Parco archeologico) 2. Grandi aree di svago, e di incontaminazione
Mental wellbeing	1. Possibilità di pensare un vivere e un lavorare/progettare più a misura d'uomo 2. Il relativo equilibrio tra natura e costruito (ma anche il costruito laddove lo permetta) ** 3. Possibilità di pianificare il proprio lavoro in maniera più efficace su quelle che sono le necessità della comunità locale (val di fiastra e comunità Dei piccoli borghi) 4. Ambiente capace di infondere serenità 5. La ridotta pressione antropica agisce favorevolmente su elementi quali congestione, i tempi di vita, l'intensità delle relazioni 6. Qualità della vita, tempi di vita lenti, silenzio 7. Potenziali effetti sulla disponibilità di equilibrio psichico 8. Qualità della vita 9. I benefici sono vivere nel silenzio ed. arrivare velocemente in una città caotica. Poter spostarsi tra un cantiere e l'altro e non smettere mai di guardarsi intorno meravigliati 10. Respiro più ampio (***). Serenità 11. Se il rapporto è virtuoso, permette di agire i propri bisogni. Restituisce il ritmo sensato e imprescindibile 12. La natura aiuta a scandire i tempi e gli spazi dell'abitare, del fare 13. Muoversi liberamente, fermarsi a riflettere. Aver tempo per noi
Intrinsic value	1. La natura è l'elemento primo e fondamentale della valle, cioè da cui esso trae la sua essenza. 2. La val di fiastra è un prodotto di un lungo processo di antropizzazione. Il concetto di natura va dunque contestualizzato. È una "costruzione sociale".
Biodiversity (*together with intrinsic value)	1. Una natura peri-Urbana in Grado di contenere grande biodiversità 2. Importanza della natura come biodiversità
Sense of place	1. Siamo una vallata, siamo nati in un luogo che per conformazione ci protegge e ci dirige velocemente dal mare alla montagna. 2. Scala Dei borghi, rapporti di intimità—Che si instaurano con i luoghi 3. La val di fiastra offre una mediazione tra le aree costiere e le più interne. Mediatore culturale e di paesaggio 4. Natura patrimonio Dei luoghi in termini di identità culturale
Sense of community (*together with sense of place)	1. Pratiche dal basso e autodeterminazione, da indagare, beneficio che si potrebbe trarre dalla "replicabilità" 2. La natura è elemento identitario (Valle del fiastra, natura nel nome che la comunità si è dato) 3. Natura come servizio principale/primario per la comunità e il territorio 4. Difendere il nostro ambiente naturale e crearne un opportunità di luogo dove vivere e lavorare sarà la sfida più importante che abbiamo davanti a noi
Aesthetic beauty	1. Viste mozzafiato 2. La bellezza quasi incontaminata ed. un paesaggio unico
Source of inspiration (*together with aesthetic beauty)	1. È cornice di attivazione progettuale, sguardo a processi partecipativi

ES Category	Post-IT Content
Artisan products	1. Il legame d'oro marchigianità—Enogastronomia (** sense of place) 2. Artigianato locale e gastronomia di ottima qualità 3. Saperi per la trasformazione Dei prodotti dell'agricoltura 4. Indagando tra gli autoctoni, filiera corta 5. Produttori legati al territorio 6. Prodotti enogastronomici locali
Agricultural products	1. I prodotti dell'agricoltura locale per la loro genuinità sostengono l'economia ed. è apprezzabile il contributo legato alla ristorazione. Molta può esser fatto ancora per creare nuove filiere 2. Natura è agricoltura e allevamento 3. Cibo 4. Sostentamento
Drinking water	1. L'economia è sostentata facilmente dalla natura che offre acqua per i campi 2. Acqua per i campi
Air purification	1. Aria
Disservices (not included in the study)	1. Laddove distrugge
Others not included within ES categories	1. Risultato di lungo periodo del rapporto comunità antropica/natura. È questa interazione che beneficia le economie locali 2. La natura è in ogni impresa 3. È la natura selvaggia, ma anche quella coltivata, quasi rigogliosa. Questo contrasto tra le colture locali e il fiume lasciato quasi al suo corso, genera al voglia di rendere più accessibile questi luoghi, ma non solo per i turisti, quanto per coloro che vi abitano. Deve diventare uno scambio equo tra società ed. habitat 4. La natura che diventa infrastruttura 5. La natura come estensione del centro urbano e non come qualcosa di altro 6. Tornare ad essere al centro del nostro sviluppo economico 7. In una realtà come la val di fiastra la natura è alla base dell'economia, in tutti i settori 8. Simbolo dell'operosità tipica delle Marche 9. Difficile immaginare una natura isolata: Il servizio benessere è dato dalla comprensione ed. il relativo equilibrio tra dato naturale e vite degli abitanti

1.5   Appendix E: Full Script of the Questionnaire

Ciao,

il questionario che segue riguarda il tuo ruolo come attore locale nella co-produzione dei servizi ecosistemici della Val di Fiastra.

Iniziamo selezionando un servizio ecosistemico in cui hai un ruolo. Il tuo ruolo potrebbe essere centrale nell'offerta del servizio, ma anche di semplice beneficiario o interessato.

1.5.2.1   Sezione A: Ruolo Nella Coproduzione di Servizio

1. Hai un ruolo in queste categorie di servizi? (Seleziona tutte le voci applicabili)

    (a) P1 Produzioni Agricole
    (b) P2 Offerta Acqua Potabile
    (c) P3 Offerta Energia Idroelettrica
    (d) R1 Regolazione Idraulica
    (e) R2 Regolazione Climatica
    (f) R3 Purificazione dell'Aria
    (g) C1 Eco-Turismo
    (h) C2 Educazione Ambientale
    (i) C3 Attività Sportiva
    (j) C4 Valore Intrinseco
    (k) C5 Senso di appartenenza alla comunità
    (l) C6 Bellezza
    (m) C7 Benessere Psico-fisico
    (n) C8 Prodotti artigianali ed. Enogastronomici

2. Stai considerando un servizio specifico all'interno della categoria?

3. Qual è il tuo ruolo?

    (a) Utente (fai esperienza di questo servizio e ne ricevi i benefici)
    (b) Manager (la tua attività permette l'offerta di questo servizio)
    (c) Influenzato negativamente (sei infastidito dalla presenza di questo servizio)
    (d) Interessato (non sei utente diretto ma credi nella sua importanza)
    (e) Investigatore (ti occupi di questo servizio da un punto di vista di ricerca)
    (f) Altro: ____

1.5.2.2   Sezione B: Capitali Antropici

4. [capitale umano] Quanto di questo ruolo è collegato alla tua conoscenza, educazione, motivazioni, abilità o salute? Ad esempio: il tuo ruolo in questo servizio è collegato al tuo diploma o a un certificato che hai ottenuto ad un corso di formazione.
   1   2   3   4   5
5. [capitale sociale] Quanto di questo ruolo è collegato a valori e norme, reti formali e informali o fiducia? Ad esempio: il tuo ruolo in questo servizio è collegato alle persone che conosci nell'ambiente professionale/associativo, la fiducia che i clienti hanno su di te
6. 1   2   3   4   5
7. [capitale fisico] Quanto di questo ruolo è collegato ai tuoi macchinari, strumenti, infrastrutture o capitale costruito? Ad esempio: il tuo ruolo in questo servizio è collegato a uno specifico macchinario che possiedi
8. 1   2   3   4   5
9. [capitale finanziario] Quanto di questo ruolo è collegato ai tuoi risparmi, crediti, sovvenzioni o pagamenti diretti? Ad esempio: il tuo ruolo in questo servizio è collegato al denaro che hai investito in esso / che utilizzi per usufruire del servizio
   1   2   3   4   5

1.5.2.3   Sezione C: Dipendenza e Relazioni Su Larga Scala

10. Quanto diresti che questo servizio è importante per te? Quanto fai affidamento/ dipendi da questo servizio (per il tuo quotidiano/per la tua vita/per la tua attività)? es: economicamente o idealmente
    1 2 3 4 5
11. Credi che questo servizio sia importante per altri attori? (fanno affidamento/ dipendono da questo servizio)

    (a) A Agricoltori
    (b) A Produttori locali
    (c) A Grande distribuzione
    (d) A Distribuzione di prossimità A Bar, ristoranti e agriturismi B Istituti scolastici
    (e) B Centri di Educazione ambientale o altre attività (Es. Mamma asina a Colmurano)

(f) B Esperti e Studiosi nei settori ambientali

(g) C Gestori Turistici

(h) C Associazioni per la promozione locale

(i) C Attori legati alla ricreazione locale (Jogging etc) C Turisti regionali

(j) C Turisti stranieri

(k) D Amministrazioni Comunali

(l) D Architetti e pianificatori

(m) D Autorità regionali

(n) E Cittadini della Val di Fiastra

(o) E Cittadini delle città (es: Macerata, Civitanova, Ancona)

(p) Altro: ____

12. Chi beneficia di questo servizio?

(a) A Agricoltori

(b) A Produttori locali

(c) A Grande distribuzione

(d) A Distribuzione di prossimità A Bar, ristoranti e agriturismi B Istituti scolastici

(e) B Centri di Educazione ambientale o altre attività (Es. Mamma asina a Colmurano)

(f) B Esperti e Studiosi nei settori ambientali

(g) C Gestori Turistici

(h) C Associazioni per la promozione locale

(i) C Attori legati alla ricreazione locale (Jogging etc) C Turisti regionali

(j) C Turisti stranieri

(k) D Amministrazioni Comunali

(l) D Architetti e pianificatori

(m) D Autorità regionali

(n) E Cittadini della Val di Fiastra

(o) E Cittadini delle città (es: Macerata, Civitanova, Ancona)

(p) Altro: ____

13. Nel caso hai un ruolo nella fornitura del servizio, con chi collabori per la fornitura di questo servizio?

(a) A Agricoltori

(b) A Produttori locali

(c) A Grande distribuzione

(d) A Distribuzione di prossimità A Bar, ristoranti e agriturismi B Istituti scolastici

(e) B Centri di Educazione ambientale o altre attività (Es. Mamma asina a Colmurano)

(f) B Esperti e Studiosi nei settori ambientali

  (g) C Gestori Turistici

  (h) C Associazioni per la promozione locale

  (i) C Attori legati alla ricreazione locale (Jogging etc) C Turisti regionali

  (j) C Turisti stranieri

  (k) D Amministrazioni Comunali

  (l) D Architetti e pianificatori

  (m) D Autorità regionali

  (n) E Cittadini della Val di Fiastra

  (o) E Cittadini delle città (es: Macerata, Civitanova, Ancona)

  (p) Altro: ____

1.5.2.4   Sezione D: Influenza Sulle Decisioni Relative al Servizio

14. Contribuisci alle decisioni relative alla gestione o allo stato di questo servizio?
 1 2 3 4 5

15. Quali altri attori contribuiscono alle decisioni relative alla gestione o allo stato di questo servizio?

  (a) A Agricoltori

  (b) A Produttori locali

  (c) A Grande distribuzione

  (d) A Distribuzione di prossimità A Bar, ristoranti e agriturismi B Istituti scolastici

  (e) B Centri di Educazione ambientale o altre attività (Es. Mamma asina a Colmurano)

  (f) B Esperti e Studiosi nei settori ambientali

  (g) C Gestori Turistici

  (h) C Associazioni per la promozione locale

  (i) C Attori legati alla ricreazione locale (Jogging etc) C Turisti regionali

  (j) C Turisti stranieri

  (k) D Amministrazioni Comunali

  (l) D Architetti e pianificatori

  (m) D Autorità regionali

  (n) E Cittadini della Val di Fiastra

  (o) E Cittadini delle città (es: Macerata, Civitanova, Ancona)

  (p) Altro: ____

16. Chi altro potresti suggerirci potrebbe avere conoscenze pertinenti su questo servizio?

1.5.2.5   Sezione E: Informazioni Generali Sull'intervistato

17. Hai visitato qualche area protetta naturale durante l'ultimo anno?

  (a) Sì
  (b) No

Se sì, dove? ____

18. Di solito acquisti o consumi prodotti biologici e/o a km0?

  1 2 3 4 5

19. Di solito separi i rifiuti?

  1 2 3 4 5

20. In che comune abiti?

____

21. Qual è il tuo genere?

  (a) Donna
  (b) Uomo
  (c) Altro ___________

22. Quanti anni hai?

____

23. Qual è la tua attività? che lavoro fai?

____

1.6   Appendix F: Characterization of Questionnaire Respondents

This chapter presents a profile of the demographic and socioeconomic characteristics of the participants of the questionnaire. Respondents were aged between 22 and 71, with a similar proportion men and women. They had a good representation of the six municipal territories (Fig. F.10).

As the environmental awareness could influence questionnaire answers, participants were asked about their environmental behavior. The results show high preference for biological or local products, while almost the whole sample makes recycling collection (Fig. F.11).

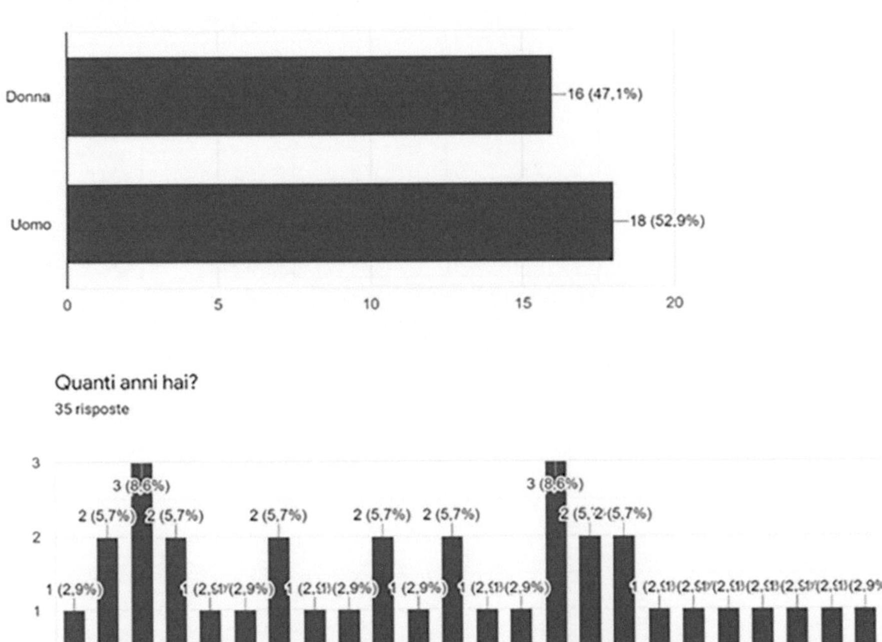

Fig. F.10 Age of the respondents (below) and number and percentage of respondents' gender (above)

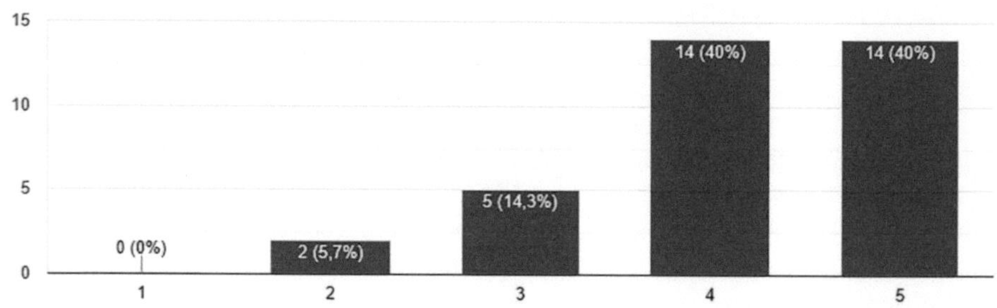

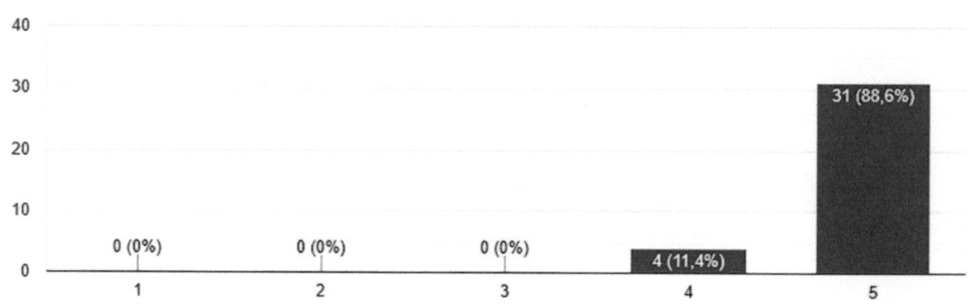

Fig. F.11 Environmental behavior of questionnaire respondents

1.7 Appendix G: Statistical Analysis

To answer the research questions, the data collected in the focus groups and the interviews were organized, statistically analyzed, and visualized through R software (RStudio Team, 2021), using packages fmsb (Nakazawa, 2021), dplyr (Wickham et al., 2021), and ggplot2 (Wickham, 2016). The full code is available at the supplementary materials (Appendix 4.E).

Within data organization, respondents' answers in questionnaire sections D and E were normalized per each stakeholder and per stakeholder groups, by dividing the number of selected stakeholders by the total number of stakeholders within the stakeholders group. The results were then aggregated per ES and stakeholder classification of the respondent.

Regarding the statistical analysis, the Shapiro-Wilk test was used to check whether the respondents' answers followed a normal distribution. As for the preferences expressed by focus groups, it showed that the distribution of the answers by participants departed significantly from normality. The test was run also to analyze the distribution of the answers of questionnaire respondents regarding the capitals involved in NCP co-production and again the result differed from normality.

Based on this outcome, non-parametric tests were used to verify the significance of the differences among the responses. As for the preferences expressed in numerical data within the focus group, Wilcoxon's rank sum test was used to assess the significance of the differences between the Focus group A: Civil Society and the Focus group B: Administration. As for the differences on capitals involved in NCP co-production, giving that the Likert scale employed (1 to 5) consists in categorical data, the Chi-square test was performed.

1.8　Appendix H: R Code Study 2

```r
mydata <- read.csv (file.choose("C:\\"))


#function Sapply and Summary give general information on
data frame
sapply(input_R,class)
summary(input_R)

#transform input data in data.frame and integer numbers
in numeric
input_R_FG <- data.frame(input_R_FG)
input_R_FG$CS <- as.numeric(input_R_FG$CS)
input_R_FG$MA <- as.numeric(input_R_FG$MA)

#install the packages from the library
library(tidyverse) library(dplyr) library(tidyr) library(ggplot2)
library(ggthemes) library(fmsb) library(gridExtra) library(grid)
library(lattice) library("xlsx")

SH_DIN %>% add_row(ES = "C1")

# related to ES
P1 <- dplyr::filter(input_R,ES=="P1")
P2 <- dplyr::filter(input_R,ES=="P2")
R1 <- dplyr::filter(input_R,ES=="R1")
R2 <- dplyr::filter(input_R,ES=="R2")
C1 <- dplyr::filter(input_R,ES=="C1")
C2 <- dplyr::filter(input_R,ES=="C2")
C3 <- dplyr::filter(input_R,ES=="C3")
C4 <- dplyr::filter(input_R,ES=="C4")
C5 <- dplyr::filter(input_R,ES=="C5")
C6 <- dplyr::filter(input_R,ES=="C6")
C7 <- dplyr::filter(input_R,ES=="C7")
C8 <- dplyr::filter(input_R,ES=="C8")

#related to SH GROUP
PRO <- dplyr::filter(input_R,SH.GROUP=="PRO")
TOU <- dplyr::filter(input_R,SH.GROUP=="TOU")
PLA <- dplyr::filter(input_R,SH.GROUP=="PLA")
SOC <- dplyr::filter(input_R,SH.GROUP=="SOC")
SCH <- dplyr::filter(input_R,SH.GROUP=="SCH")
```

```r
temp <- input_R %>%
  group_by(ES, Role, SH.GROUP) %>%
  summarise_each(funs(mean))
Role_ES <- subset(temp, select = -c(SH, ï..ID, IMP, BEN, CON,
                                    Contribution,
                                    HC, SC, PC, FC,
                                    IMP_PRO,     IMP_SCH,     IMP_
TOU,     IMP_PLA,     IMP_SOC,    IMP_TOT,
                                    BEN_PRO,     BEN_SCH,     BEN_
TOU,     BEN_PLA,     BEN_SOC,    BEN_TOT,
                                    CON_PRO,     CON_SCH,     CON_
TOU,     CON_PLA,     CON_SOC,    CON_TOT))
Role_ES$SH.GROUP_Role <- paste(Role_ES$SH.GROUP,Role_
ES$Role,sep=" - ")

temp <- input_R %>%
  group_by(ES) %>%
  summarise_each(funs(mean()))
temp2 <- subset(temp, select = -c(SH,Role, ï..ID,
                                  SH.GROUP, IMP, BEN, CON,
                                  Importance, Contribution))
CAP_ES <- subset(temp2, select = -c(IMP_PRO,     IMP_SCH,     IMP_
TOU,     IMP_PLA,     IMP_SOC,    IMP_TOT,
                                    BEN_PRO,     BEN_SCH,     BEN_
TOU,     BEN_PLA,     BEN_SOC,    BEN_TOT,
                                    CON_PRO,     CON_SCH,     CON_
TOU,     CON_PLA,     CON_SOC,    CON_TOT))

temp <- input_R %>%
  group_by(ES, SH.GROUP) %>%
  summarise_each(funs(mean))
SH_DIN <- subset(temp, select = -c(SH,Role, ï..ID, IMP, BEN, CON,
                                   HC, SC, PC, FC))

temp <- input_R %>%
  group_by(ES) %>%
  summarise_each(funs(mean))
SH_DIN_TOT <- subset(temp, select = -c(SH, SH.GROUP, Role, ï..ID,
IMP, BEN, CON,
                                       HC, SC, PC, FC))

temp <- input_R_COL %>%
  group_by(SH.GROUP) %>%
  summarise_each(funs(mean))
SH.GROUP_COL <- subset(temp, select = -c(ES, SH, Role,
COL, ï..ID))
```

```
temp <- input_R_COL %>%
  group_by(SH) %>%
  summarise_each(funs(mean))
SH_COL <- subset(temp, select = -c(ES, SH.GROUP, Role,
COL, ï..ID))
```

```
FG1 <- dplyr::filter(input_R_FG,FG=="CS")
FG2 <- dplyr::filter(input_R_FG,FG=="MA")
```

```
SH_DIN_ES <-
  SH_DIN %>%
  group_by(ES) %>%
  summarise_each(funs(mean))
```

```
SH_DIN_SH.GROUP <-
  SH_DIN %>%
  group_by(SH.GROUP) %>%
  summarise_each(funs(mean))
```

```
# the p-value > 0.05 implying that the distribution of
# the data are not significantly different from normal distribution
shapiro.test(input_R$HC)
shapiro.test(input_R$SC)
shapiro.test(input_R$PC)
shapiro.test(input_R$FC)

shapiro.test(input_R_FG$VOTE)
# results W = 0.9058, p-value = 0.01569
```

```
# in case the data are categorical // chi-square test
# in case the data are numerical // Kruskal test
```

```
# For a Chi-square test, a p-value that is less than or equal to your
# significance level indicates there is sufficient evidence to
conclude
# that the observed distribution is not the same as the expected
distribution.
# You can conclude that a relationship exists between the
categorical variables.
```

```
# is there significant difference among SH.GROUP for the value they
associate to capitals?
chisq.test(table(input_R$SH.GROUP,input_R$HC))
chisq.test(table(input_R$SH.GROUP,input_R$SC))
chisq.test(table(input_R$SH.GROUP,input_R$PC))
chisq.test(table(input_R$SH.GROUP,input_R$FC))
#no significant difference for any Capital
```

```
#is there significant difference among ES for the value the
respondents associate to capitals?
chisq.test(table(input_R$ES,input_R$HC))
chisq.test(table(input_R$ES,input_R$SC))
chisq.test(table(input_R$ES,input_R$PC))
chisq.test(table(input_R$ES,input_R$FC))
#significant difference only for social capital (SC) - p-value
= 0.03268
```

 # P-value ≤ α: The differences between some of the medians are statistically significant
```
# P-value > α: The differences between the medians are not
statistically significant

#is there significant difference among SH.GROUP for the value they
associate to capitals?
kruskal.test(HC ~ SH.GROUP, data = input_R)
kruskal.test(SC ~ SH.GROUP, data = input_R)
kruskal.test(PC ~ SH.GROUP, data = input_R)
kruskal.test(FC ~ SH.GROUP, data = input_R)
#result: there is no relevant relationship

#is there significant difference among ES for the value the
respondents associate to capitals?
kruskal.test(HC ~ ES, data = input_R)
kruskal.test(SC ~ ES, data = input_R)
kruskal.test(PC ~ ES, data = input_R)
kruskal.test(FC ~ ES, data = input_R)
#result: there is no relevant relationship
```

wilcox.test(FG1$VOTE, FG2$VOTE)

 # for the four main ES

```
# C5 Sense of place
proportions <- table(C5$Role)/length(C5$Role)
percentages <- proportions*100
view(percentages)

# C1 Eco-Tourism
proportions <- table(C1$Role)/length(C1$Role)
percentages <- proportions*100
view(percentages)

# C8 Artisan products
proportions <- table(C8$Role)/length(C8$Role)
percentages <- proportions*100
view(percentages)
```

```r
# P1 Agricultural products
proportions <- table(P1$Role)/length(P1$Role)
percentages <- proportions*100
view(percentages)
```

#plot ES-SH, with horizontal bars

```r
input_R_FG %>%
  ggplot(aes(x=ï.., y=VOTE))+
  geom_col(aes(fill=FG),
           width = 0.3,
           alpha=0.5)+
  labs(y="Nr. Preferences", x="ES", title = "Preferences for
Ecosystem Services")+
  coord_flip()+
  geom_vline(xintercept=11.5, colour="black")+
  geom_vline(xintercept=8.5, colour="black")+
  theme_bw()
```

#input_R %>%

```r
#   ggplot(aes(ES, SH))+
#   geom_point(aes(size=Importance,
#                  colour=Role,
#                  alpha = 0.5))+
#   labs(title = "Role of local stakeholders in ES
co-production")+
#   theme_bw()+
#   geom_hline(yintercept=4.5, colour="black")+
#   geom_hline(yintercept=7.5, colour="black")+
#   geom_hline(yintercept=11.5, colour="black")+
#   geom_hline(yintercept=16.5, colour="black")
```

Role_ES %>%

```r
  ggplot(aes(ES, SH.GROUP_Role))+
  geom_point(aes(size=Importance,
                 colour=Role,
                 alpha = 0.5))+
  labs(title = "Role of local stakeholders in ES co-production",
       y="Stakeholders groups")+
  theme_bw()+
  geom_hline(yintercept=3.5, colour="black")+
  geom_hline(yintercept=5.5, colour="black")+
  geom_hline(yintercept=8.5, colour="black")+
  geom_hline(yintercept=11.5, colour="black")
```

 # name rows as the first column (as asked by radarchart funct)
```r
CAP_ES <- CAP_ES %>% remove_rownames %>% column_to_
rownames(var="ES")

# I divide the data in four (Select rows by position) and add max
and min values (as asked by radarchart funct)
temp <- dplyr::slice(CAP_ES, 1:4)
CAP_ES1 <- data.frame(rbind(rep(5,1), rep(1,1), temp))
temp <- dplyr::slice(CAP_ES, 4:8)
CAP_ES2 <- data.frame(rbind(rep(5,1), rep(1,1), temp))
temp <- dplyr::slice(CAP_ES, 9:10)
CAP_ES3 <- data.frame(rbind(rep(5,1), rep(1,1), temp))
temp <- dplyr::slice(CAP_ES, 11:12)
CAP_ES4 <- data.frame(rbind(rep(5,1), rep(1,1), temp))

#I use the function par to aggregate the 4 graphs
op <- par(mar=c(1, 2, 2, 1),mfrow=c(2, 2))

# I build radar 1
CAP_ES1 %>%
  radarchart(cglty = 1,            # Grid line type
             cglcol = "black",     # Grid line color
             pcol = 1:4,           # Color for each line
             plwd = 2,             # Width for each line
             plty = 1,             # Line type for each line
             pty = 3,
             axistype = 1,
             caxislabels = c("1","2", "3", "4", "5"),
             title = "Anthropogenic Capitals involved in ES
co-production")+
  legend("topright",
         legend=(c("C1 Eco-turism","C2 Env. education","C3 Sport
activity","C4 Intrinsic value")),
         bty= "n", pch=16,
         col=1:11,
         text.col="darkgrey", cex=1, pt.cex=1)

# I repeat the same code for building radar 2, radar 3, radar 4
```

#Group "Production"

```r
g1 <-
SH_DIN %>%
  group_by(ES) %>%
  summarise_each(funs(mean)) %>%
  ggplot()+
  geom_col(aes(x=IMP_PRO,y=ES),
           width = 0.12,
           alpha=0.5,
           position = position_nudge(y = -0.15),
           fill = "blue",
           colour ="blue")+
  geom_col(aes(x=BEN_PRO,y=ES),
           width = 0.12,
           alpha=0.5,
           position = position_nudge(y = 0),
           fill="red",
           col="red")+
  geom_col(aes(x=CON_PRO,y=ES),
           width = 0.12,
           alpha=0.5,
           position = position_nudge(y = 0.15),
           fill="grey",
           col="black")+
  labs(title = "Production", x="Avg. responses")+
  lims(x=c(0,1))+
  theme_bw()

# I repeat the same code for Group "school and research", Group
"Tourism and commerce", #Group "Planning and administration",
#Group "Society"


# Total (the five groups together)
grid.arrange(g1, g2, g3, g4, g5, nrow=2)
```

```r
#Group "Production"
#g1b <-
#temp <- data=filter(SH_DIN, SH.GROUP=="PRO")
input_R %>%
#   filter(SH.GROUP=="PRO")%>%
  ggplot()+
  geom_boxplot((aes(y=Contribution,x=ES)),
                 size = 1,
                 alpha=0.5)+
  labs(title = "Production", x="Self responses",
y="Decision-making")+
lims(y=c(0,5))+
  theme_bw()
  labs(title = "Production (self)", x="Avg. responses")

# I repeat the same code for Group "school and research", Group
"Tourism and commerce", #Group "Planning and administration",
#Group "Society"


#Total (the five groups together)
grid.arrange(g1b, g2b, g3b, g4b, g5b, nrow=2)

LOCATION <- "C:/Users/toma/documents/R-output.xlsx"


write.xlsx(as.data.frame(PRO), LOCATION, sheetName = "PRO")
write.xlsx(as.data.frame(SCH), LOCATION, sheetName = "SCH",
append=TRUE)
write.xlsx(as.data.frame(TOU), LOCATION, sheetName = "TOU",
append=TRUE)

write.xlsx(as.data.frame(PLA), LOCATION, sheetName = "PLA",
append=TRUE)
write.xlsx(as.data.frame(SOC), LOCATION, sheetName = "SOC",
append=TRUE)
write.xlsx(as.data.frame(SH_DIN), LOCATION, sheetName = "SH_DIN",
append=TRUE)
write.xlsx(as.data.frame(SH_DIN_ES), LOCATION, sheetName =
"SH_DIN_ES", append=TRUE)
write.xlsx(as.data.frame(SH_DIN_SH.GROUP), LOCATION, sheetName =
"SH_DIN_SH.GROUP", append=TRUE)
write.xlsx(as.data.frame(CAP_ES), LOCATION, sheetName = "CAP_ES",
append=TRUE)
```

References

Albert C, Aronson J, Fürst C, Opdam P (2014) Integrating ecosystem services in landscape planning: requirements, approaches, and impacts. Landsc Ecol 29:1277–1285. https://doi.org/10.1007/s10980-014-0085-0

Albert C, Galler C, Hermes J et al (2016) Applying ecosystem services indicators in landscape planning and management: the ES-in-planning framework. Ecol Indic 61:100–113. https://doi.org/10.1016/j.ecolind.2015.03.029

Antognelli S, Vizzari M (2017) Landscape liveability spatial assessment integrating ecosystem and urban services with their perceived importance by stakeholders. Ecol Indic 72:703–725. https://doi.org/10.1016/j.ecolind.2016.08.015

Antrop M (2005) Why landscapes of the past are important for the future. Landsc Urban Plan 70:21–34. https://doi.org/10.1016/j.landurbplan.2003.10.002

Arcidiacono A, Ronchi S, Salata S (2016) Managing multiple ecosystem Services for Landscape Conservation: a green infrastructure in Lombardy region. Proc Eng 161:2297–2303. https://doi.org/10.1016/j.proeng.2016.08.831

Aretano R, Petrosillo I, Zaccarelli N et al (2013) People perception of landscape change effects on ecosystem services in small Mediterranean islands: a combination of subjective and objective assessments. Landsc Urban Plan 112:63–73. https://doi.org/10.1016/j.landurbplan.2012.12.010

Arias-Arévalo P, Martín-López B, Gómez-Baggethun E (2017) Exploring intrinsic, instrumental, and relational values for sustainable management of social-ecological systems. E&S 22:art43. https://doi.org/10.5751/ES-09812-220443

Arnstein SR (1969) A ladder of citizen participation. J Am Inst Plann 35:216–224. https://doi.org/10.1080/01944366908977225

Artmann M, Bastian O, Grunewald K (2017) Using the concepts of green infrastructure and ecosystem services to specify Leitbilder for compact and green cities—the example of the landscape plan of Dresden (Germany). Sustainability 9:198. https://doi.org/10.3390/su9020198

Artmann M, Inostroza L, Fan P (2019) Urban sprawl, compact urban development and green cities. How much do we know, how much do we agree? Ecol Indic 96:3–9. https://doi.org/10.1016/j.ecolind.2018.10.059

Balzan MV, Caruana J, Zammit A (2018) Assessing the capacity and flow of ecosystem services in multifunctional landscapes: evidence of a rural-urban gradient in a Mediterranean small Island state. Land Use Policy 75:711–725. https://doi.org/10.1016/j.landusepol.2017.08.025

Balzan MV, Sadula R, Scalvenzi L (2020) Assessing ecosystem services supplied by agroecosystems in Mediterranean Europe: a literature review. Land 9:245. https://doi.org/10.3390/land9080245

Barca F (2009) An agenda for a reformed cohesion policy: a place-based approach to meeting European Union challenges and expectations (Barca report). European Parlament, Bruxelles

Barca F (2018) Politica di coesione: tre mosse. Documenti IAI

Barca F (2022) Introduzione. In: L'Italia lontana: Una politica per le aree interne. Donzelli Editore

Barca F, Casavola P, Lucatelli S (2014) Strategia nazionale per le Aree interne: definizione, obiettivi, strumenti e governance

Barca F, McCann P, Rodríguez-Pose A (2012) The case for regional development intervention: place-based versus place-neutral approaches*. J Reg Sci 52:134–152. https://doi.org/10.1111/j.1467-9787.2011.00756.x

Baró F, Gómez-Baggethun E, Haase D (2017) Ecosystem service bundles along the urban-rural gradient: insights for landscape planning and management. Ecosyst Serv 24:147–159. https://doi.org/10.1016/j.ecoser.2017.02.021

Beckmann-Wübbelt A, Fricke A, Sebesvari Z et al (2021) High public appreciation for the cultural ecosystem services of urban and peri-urban forests during the COVID-19 pandemic. Sustain Cities Soc 74:103240. https://doi.org/10.1016/j.scs.2021.103240

Bennett EM, Cramer W, Begossi A et al (2015) Linking biodiversity, ecosystem services, and human Well-being: three challenges for designing research for sustainability. Curr Opin Environ Sustain 14:76–85. https://doi.org/10.1016/j.cosust.2015.03.007

Bennett EM, Baird J, Baulch H et al (2021) Ecosystem services and the resilience of agricultural landscapes. In: Bohan DA, Vanbergen AJ (eds) Advances in ecological research. Academic Press, pp 1–43

Benra F, Nahuelhual L, Felipe-Lucia M et al (2022) Balancing ecological and social goals in PES design—single objective strategies are not sufficient. Ecosyst Serv 53:101385. https://doi.org/10.1016/j.ecoser.2021.101385

Bevilacqua P (2013) Marche. In: Agnoletti M (ed) Italian historical rural landscapes: cultural values for the environment and rural development. Springer, Dordrecht, pp 343–361

Binder C, Hinkel J, Bots P, Pahl-Wostl C (2013) Comparison of frameworks for analyzing social-ecological systems. Ecol Soc 18:19. https://doi.org/10.5751/ES-05551-180426

Blondel J (2006) The 'design' of Mediterranean landscapes: a millennial story of humans and ecological systems during the historic period. Hum Ecol 34:713–729. https://doi.org/10.1007/s10745-006-9030-4

Bobbio L, Pomatto G (2007) Il coinvolgimento dei cittadini nelle scelte pubbliche. Meridian, pp 45–67

Brunner SH, Huber R, Grêt-Regamey A (2017) Mapping uncertainties in the future provision of ecosystem services in a mountain region in Switzerland. Reg Environ Chang 17:2309–2321. https://doi.org/10.1007/s10113-017-1118-4

Bruno D, Sorando R, Álvarez-Farizo B et al (2021) Depopulation impacts on ecosystem services in Mediterranean rural areas. Ecosyst Serv 52:101369. https://doi.org/10.1016/j.ecoser.2021.101369

Burkhard B, Kroll F, Nedkov S, Müller F (2012) Mapping ecosystem service supply, demand and budgets. Ecol Indic 21:17–29. https://doi.org/10.1016/j.ecolind.2011.06.019

Burkhard B, Kandziora M, Hou Y, Müller F (2014) Ecosystem service potentials, flows and demands-concepts for spatial localisation, indication and quantification. Landscape Online 34:1. https://doi.org/10.3097/LO.201434

Calafati A (2015) Città tra sviluppo e declino. Donzelli Editore, [S.l.]

Calafati A, Mazzoni F (2009) Città in nuce nelle Marche. Coalescenza territoriale e sviluppo economico. Franco Angeli

Calderón-Argelich A, Benetti S, Anguelovski I et al (2021) Tracing and building up environmental justice considerations in the urban ecosystem service literature: a systematic review. Landsc Urban Plan 214:104130. https://doi.org/10.1016/j.landurbplan.2021.104130

Camagni R (2008) Regional competitiveness: towards a concept of territorial capital. In: Capello R, Camagni R (eds) Handbook of regional growth and development theories. Edward Elgar, Cheltenham, pp 239–256

Carmona-Torres C, Parra-López C, Groot JCJ, Rossing WAH (2011) Collective action for multi-scale environmental management: achieving landscape policy objectives through cooperation of local resource managers. Landsc Urban Plan 103:24–33. https://doi.org/10.1016/j.landurbplan.2011.05.009

Chamusca P (2023) From ecological marginality to spatial centrality: rethinking cohesion through environmental dynamics. J Territorial Dev Stud 12(1):45–61

Chan KMA, Satterfield T, Goldstein J (2012) Rethinking ecosystem services to better address and navigate cultural values. Ecological Economics 74:8–18. https://doi.org/10.1016/j.ecolecon.2011.11.011

Chape S, Harrison J, Spalding M, Lysenko I (2005) Measuring the extent and effectiveness of protected areas as an indicator for meeting global biodiversity targets. Philosophical Transactions of the Royal Society B: Biological Sciences 360:443–455. https://doi.org/10.1098/rstb.2004.1592

Chen C (2006) CiteSpace II: detecting and visualizing emerging trends and transient patterns in scientific literature. J Am Soc Inf Sci 57:359–377. https://doi.org/10.1002/asi.20317

Chen X, de Vries S, Assmuth T et al (2019) Research challenges for cultural ecosystem services and public health in (peri-)urban environments. Sci Total Environ 651:2118–2129. https://doi.org/10.1016/j.scitotenv.2018.09.030

Collettivo PRiNT (2022) Aree interne e comunità: Cronache dal cuore dell'Italia

Costanza R, de Groot R, Sutton P et al (2014) Changes in the global value of ecosystem services. Glob Environ Chang 26:152–158. https://doi.org/10.1016/j.gloenvcha.2014.04.002

Council of Europe (2000) The European Landscape Convention (Florence, 2000) European Treaty Series - No 176

Crossman ND, Burkhard B, Nedkov S et al (2013) A blueprint for mapping and modelling ecosystem services. Ecosyst Serv 4:4–14. https://doi.org/10.1016/j.ecoser.2013.02.001

Cumming GS, Allen CR (2017) Protected areas as social-ecological systems: perspectives from resilience and social-ecological systems theory. Ecol Appl 27:1709–1717. https://doi.org/10.1002/eap.1584

Cumming GS, Buerkert A, Hoffmann EM et al (2014) Implications of agricultural transitions and urbanization for ecosystem services. Nature 515:50–57. https://doi.org/10.1038/nature13945

de Groot R (2006) Function-analysis and valuation as a tool to assess land use conflicts in planning for sustainable, multi-functional landscapes. Landsc Urban Plan 75:175–186. https://doi.org/10.1016/j.landurbplan.2005.02.016

de Groot R, Costanza R, Braat L, Brander L, Burkhard B, Carrascosa JL, Crossman ND, Egoh BN, Geneletti D, Hansjuergens B, Hein L, Jacobs S, Kubiszewski I, Leimona B, Li B, Liu J, Luque S, Maes J, Marais C, … Willemen L (2018) Ecosystem Services are Nature's Contributions to People: Response to: Assessing nature's contributions to people. Science Progress 359(6373). https://research.utwente.nl/en/publications/ecosystem-services-are-natures-contributions-to-people-response-t

de Groot RS, Alkemade R, Braat L et al (2010a) Challenges in integrating the concept of ecosystem services and values in landscape planning, management and decision making. Ecol Complex 7:260–272. https://doi.org/10.1016/j.ecocom.2009.10.006

de Groot RS, Fisher B, Christie M et al (2010b) Integrating the ecological and economic dimensions in biodiversity and ecosystem service valuation. In: The Economics of Ecosystems and Biodiversity (TEEB). Ecological and Economic Foundations, pp 9–40

de Vos A, Biggs R, Preiser R (2019) Methods for understanding social-ecological systems: a review of place-based studies. Ecol Soc 24:16. https://doi.org/10.5751/ES-11236-240416

Derks J, Giessen L, Winkel G (2020) COVID-19-induced visitor boom reveals the importance of forests as critical infrastructure. Forest Policy Econ 118:102253. https://doi.org/10.1016/j.forpol.2020.102253

Díaz S, Demissew S, Carabias J et al (2015) The IPBES conceptual framework — connecting nature and people. Curr Opin Environ Sustain 14:1–16. https://doi.org/10.1016/j.cosust.2014.11.002

Díaz S, Pascual U, Stenseke M et al (2018) Assessing nature's contributions to people. Science 359:270–272. https://doi.org/10.1126/science.aap8826

Dick J, Turkelboom F, Woods H et al (2018) Stakeholders' perspectives on the operationalisation of the ecosystem service concept: results from 27 case studies. Ecosyst Serv 29:552–565. https://doi.org/10.1016/j.ecoser.2017.09.015

Dong S, Wen L, Liu S et al (2011) Vulnerability of worldwide pastoralism to global changes and interdisciplinary strategies for sustainable pastoralism. Ecol Soc 16:23. https://doi.org/10.5751/ES-04093-160210

EdT (2018) Sul fronte del sisma. Un'inchiesta militante sul post-terremoto dell'Appennino centrale (2016–2017)—Emidio di Treviri. Derive Approdi—Doc(k)s

Elbakidze M, Angelstam P, Yamelynets T et al (2017) A bottom-up approach to map land covers as potential green infrastructure hubs for human Well-being in rural settings: a case study from Sweden. Landsc Urban Plan 168:72–83. https://doi.org/10.1016/j.landurbplan.2017.09.031

Elbakidze M, Gebrehiwot M, Angelstam P et al (2018) Defining priority land covers that secure the livelihoods of urban and rural people in Ethiopia: a case study based on citizens' preferences. Sustainability 10:1701. https://doi.org/10.3390/su10061701

ESPON (2021) Territorial evidence and policy advice for the prosperous future of rural areas: contribution to the long-term vision for rural areas. ESPON

Estrella R, Cattrysse D, Van Orshoven J (2014) Comparison of three ideal point-based multicriteria decision methods for afforestation planning. Forests 5:3222–3240. https://doi.org/10.3390/f5123222

European Commission (1999) European spatial development perspective (ESDP): towards balanced and sustainable development of the territory of the EU. Office for Official Publications of the European Communities, Luxembourg

European Commission (2008) Green paper on territorial cohesion: turning territorial diversity into strength. Commission of the European Communities, Brussels. COM(2008) 616 final

European Commission (2021) Long-term vision for rural areas. European Commission - European Commission. https://ec.europa.eu/commission/presscorner/detail/en/IP_21_3162. Accessed 15 Jun 2022

European Union (2007) Treaty of Lisbon amending the treaty on European Union and the treaty establishing the European Community. Official Journal of the European Union, C 306. Accessed 17 Dec 2007

Faludi A (2004) Territorial cohesion: old (French) wine in new bottles? Urban Stud 41(7):1349–1365. https://doi.org/10.1080/0042098042000223645

Faludi A (ed) (2007) Territorial cohesion and the European model of society, Illustrated edition. Lincoln Institute of Land Policy, Cambridge, Mass

Felipe-Lucia M, Comín F, Bennett E (2014) Interactions among ecosystem services across land uses in a floodplain agroecosystem. Ecol Soc 19:24. https://doi.org/10.5751/ES-06249-190120

Felipe-Lucia MR, Martín-López B, Lavorel S et al (2015) Ecosystem services flows: why stakeholders' power relationships matter. PLoS One 10:e0132232. https://doi.org/10.1371/journal.pone.0132232

Felipe-Lucia MR, Soliveres S, Penone C et al (2018) Multiple forest attributes underpin the supply of multiple ecosystem services. Nat Commun 9:4839. https://doi.org/10.1038/s41467-018-07082-4

Felipe-Lucia MR, Soliveres S, Penone C et al (2020) Land-use intensity alters networks between biodiversity, ecosystem functions, and services. Proc Natl Acad Sci 117:28140–28149. https://doi.org/10.1073/pnas.2016210117

Fischer A, Eastwood A (2016) Coproduction of ecosystem services as human–nature interactions—an analytical framework. Land Use Policy 52:41–50. https://doi.org/10.1016/j.landusepol.2015.12.004

Fischer J, Gardner TA, Bennett EM et al (2015) Advancing sustainability through mainstreaming a social–ecological systems perspective. Curr Opin Environ Sustain 14:144–149. https://doi.org/10.1016/j.cosust.2015.06.002

Folke C, Jansson Å, Larsson J, Costanza R (1997) Ecosystem appropriation by cities. Ambio 26:167–172

Foray D, David PA, Hall B (2009) Smart specialization—the concept. Knowledge economists policy brief, no. 9. European Commission

Forman RTT, Godron M (1986) Landscape ecology. Wiley, New York

France-Presse A (2022) Italy declares state of emergency in drought-hit northern regions. The Guardian

Friedrich LA, Glegg G, Fletcher S et al (2020) Using ecosystem service assessments to support participatory marine spatial planning. Ocean Coastal Manag 188:105121. https://doi.org/10.1016/j.ocecoaman.2020.105121

Früh-Müller A, Hotes S, Breuer L et al (2016) Regional patterns of ecosystem services in cultural landscapes. Land 5:17. https://doi.org/10.3390/land5020017

Galli M, Berti G, Bonari E, Tatania A (2013) Manuale di progettazione partecipata per lo sviluppo sostenibile dei territori rurali. Edizioni ETS, Pisa

Gambino R (1996) Progetti per l'ambiente. Franco Angeli, Milano

García-Llorente M, Martín-López B, Iniesta-Arandia I et al (2012) The role of multi-functionality in social preferences toward semi-arid rural landscapes: an ecosystem service approach. Environ Sci Pol 19–20:136–146. https://doi.org/10.1016/j.envsci.2012.01.006

Gebre T, Gebremedhin B (2019) The mutual benefits of promoting rural-urban interdependence through linked ecosystem services. Global Ecol Conserv 20:e00707. https://doi.org/10.1016/j.gecco.2019.e00707

Geneletti D (2013) Assessing the impact of alternative land-use zoning policies on future ecosystem services. Environ Impact Assess Rev 40:25–35. https://doi.org/10.1016/j.eiar.2012.12.003

Geneletti D, Esmail BA, Cortinovis C et al (2020) Ecosystem services mapping and assessment for policy-and decision-making: lessons learned from a comparative analysis of European case studies. One Ecosystem 5:e53111. https://doi.org/10.3897/oneeco.5.e53111

Giacomelli M, Calcagni F (2022) Borgofuturo+. Un progetto locale per le aree interne, Quodlibet

Giedych R, Maksymiuk G (2017) Specific features of parks and their impact on regulation and cultural ecosystem services provision in Warsaw, Poland. Sustainability 9:792. https://doi.org/10.3390/su9050792

Gómez-Baggethun E, de Groot R, Lomas PL, Montes C (2010) The history of ecosystem services in economic theory and practice: from early notions to markets and payment schemes. Ecol Econ 69:1209–1218. https://doi.org/10.1016/j.ecolecon.2009.11.007

Gössling S (2002) Global environmental consequences of tourism. Glob Environ Chang 12:283–302. https://doi.org/10.1016/S0959-3780(02)00044-4

Grêt-Regamey A, Weibel B, Bagstad KJ et al (2014) On the effects of scale for ecosystem services mapping. PLoS One 9:e112601. https://doi.org/10.1371/journal.pone.0112601

Grêt-Regamey A, Altwegg J, Sirén EA et al (2017) Integrating ecosystem services into spatial planning—a spatial decision support tool. Landsc Urban Plan 165:206–219. https://doi.org/10.1016/j.landurbplan.2016.05.003

Haines-Young R, Potschin-Young M (2010) The links between biodiversity, ecosystem services and human well-being. In: Raffaelli DG, Frid CLJ (eds) Ecosystem ecology, 1st edn. Cambridge University Press, pp 110–139

Haines-Young R, Potschin-Young MB (2018) Revision of the common international classification for ecosystem services (CICES V5.1): A Policy Brief

Haraway D (2015) Anthropocene, Capitalocene, Plantationocene, Chthulucene: making kin. Environ Huma 6:159–165. https://doi.org/10.1215/22011919-3615934

Hill R, Díaz S, Pascual U et al (2021) Nature's contributions to people: weaving plural perspectives. One Earth 4:910–915. https://doi.org/10.1016/j.oneear.2021.06.009

Holling CS, Meffe GK (1996) Command and control and the pathology of natural resource management. Conserv Biol 10:328–337

Iniesta-Arandia I, del Amo DG, García-Nieto AP et al (2015) Factors influencing local ecological knowledge maintenance in Mediterranean watersheds: insights for environmental policies. Ambio 44:285–296. https://doi.org/10.1007/s13280-014-0556-1

IPBES (2019) Summary for policymakers of the global assessment report on biodiversity and ecosystem services. Zenodo. https://zenodo.org/record/3553579

Jericó-Daminello C, Schröter B, Mancilla Garcia M, Albert C (2021) Exploring perceptions of stakeholder roles in ecosystem services coproduction. Ecosyst Serv 51:101353. https://doi.org/10.1016/j.ecoser.2021.101353

Joint Research Centre, Institute for Environment and Sustainability, Zulian G et al (2014) ESTIMAP: ecosystem services mapping at European scale. Publications Office

Kadykalo AN, López-Rodriguez MD, Ainscough J et al (2019) Disentangling 'ecosystem services' and 'nature's contributions to people. Ecosystems and People 15:269–287. https://doi.org/10.1080/26395916.2019.1669713

Kassambara A, Mundt F (2020) Factoextra: extract and visualize the results of multivariate data analyses. In: R Package Version 1.0.7

Keeler BL, Polasky S, Brauman KA et al (2012) Linking water quality and Well-being for improved assessment and valuation of ecosystem services. Proc Natl Acad Sci 109:18619–18624. https://doi.org/10.1073/pnas.1215991109

Kopperoinen L, Itkonen P, Niemelä J (2014) Using expert knowledge in combining green infrastructure and ecosystem services in land use planning: an insight into a new place-based methodology. Landsc Ecol 29:1361–1375. https://doi.org/10.1007/s10980-014-0014-2

Krasny ME, Russ A, Tidball KG, Elmqvist T (2014) Civic ecology practices: participatory approaches to generating and measuring ecosystem services in cities. Ecosyst Serv 7:177–186. https://doi.org/10.1016/j.ecoser.2013.11.002

Krätli S, Huelsebusch C, Brooks S, Kaufmann B (2013) Pastoralism: a critical asset for food security under global climate change. Anim Front 3:42–50. https://doi.org/10.2527/af.2013-0007

Langemeyer J, Gómez-Baggethun E, Haase D et al (2016) Bridging the gap between ecosystem service assessments and land-use planning through multi-criteria decision analysis (MCDA). Environ Sci Pol 62:45–56. https://doi.org/10.1016/j.envsci.2016.02.013

Langemeyer J, Calcagni F, Baró F (2018) Mapping the intangible: using geolocated social media data to examine landscape aesthetics. Land Use Policy 77:542–552. https://doi.org/10.1016/j.landusepol.2018.05.049

Lanzas M, Hermoso V, De-Miguel S et al (2019) Designing a network of green infrastructure to enhance the conservation value of protected areas and maintain ecosystem services. Sci Total Environ 651:541–550. https://doi.org/10.1016/j.scitotenv.2018.09.164

Lavorel S, Locatelli B, Colloff MJ, Bruley E (2020) Co-producing ecosystem services for adapting to climate change. Philos Trans R Soc B 375:20190119. https://doi.org/10.1098/rstb.2019.0119

Lele S, Brondízio E, Byrne J et al (2018) Framing the environment. In: Rethinking environmentalism: linking justice, sustainability, and diversity. MIT Press, pp 1–22

Levrel H, Cabral P, Feger C et al (2017) How to overcome the implementation gap in ecosystem services? A user-friendly and inclusive tool for improved urban management. Land Use Policy 68:574–584. https://doi.org/10.1016/j.landusepol.2017.07.037

Lewis GJ, Harvey B (2001) Perceived environmental uncertainty: the extension of miller's scale to the natural environment. J Manag Stud 38:201–234. https://doi.org/10.1111/1467-6486.00234

Liu Y, Li T, Zhao W et al (2019) Landscape functional zoning at a county level based on ecosystem services bundle: methods comparison and management indication. J Environ Manag 249:109315. https://doi.org/10.1016/j.jenvman.2019.109315

Longato D, Cortinovis C, Albert C, Geneletti D (2021) Practical applications of ecosystem services in spatial planning: lessons learned from a systematic literature review. Environ Sci Pol 119:72–84. https://doi.org/10.1016/j.envsci.2021.02.001

Lucatelli S, Monaco F, Tantillo F (2019) La Strategia delle aree interne al servizio di un nuovo modello di sviluppo locale per l'Italia. Rivista economica del Mezzogiorno, pp 739–771. https://doi.org/10.1432/96257

Lucatelli S, Luisi D, Tantillo F, l'Italia AR (2022) L'Italia lontana: Una politica per le aree interne. Donzelli Editore

Magnaghi A (2010) Il progetto locale. Verso la coscienza di luogo. Bollati Boringhieri. Torino

Marino PG, Schirpke U et al (2014) Assessment and governance of ecosystem services for improving management effectiveness of Natura 2000 sites. BAE 3:229. https://doi.org/10.13128/BAE-15087

Martínez-Fernández C, Ranieri A, Sharpe S (2015) Territorial approaches for sustainable development: potential of territorial approaches for integrated and inclusive rural transformation. OECD and UNDP Working Paper

Martín-López B, Gómez-Baggethun E, García-Llorente M, Montes C (2014) Trade-offs across value-domains in ecosystem services assessment. Ecol Indic 37:220–228. https://doi.org/10.1016/j.ecolind.2013.03.003

Martín-López B, Felipe-Lucia MR, Bennett EM et al (2019) A novel telecoupling framework to assess social relations across spatial scales for ecosystem services research. J Environ Manag 241:251–263. https://doi.org/10.1016/j.jenvman.2019.04.029

Mascarenhas A, Ramos TB, Haase D, Santos R (2014) Integration of ecosystem services in spatial planning: a survey on regional planners' views. Landsc Ecol 29:1287–1300. https://doi.org/10.1007/s10980-014-0012-4

Mathey J, Rößler S, Banse J et al (2015) Brownfields as an element of green infrastructure for implementing ecosystem services into urban areas. J Urban Plann Dev 141:A4015001. https://doi.org/10.1061/(ASCE)UP.1943-5444.0000275

McCann P, Ortega-Argilés R (2015) Smart specialization, regional growth and applications to EU cohesion policy. Reg Stud 49(8):1291–1302. https://doi.org/10.1080/00343404.2013.799769

MEA (2005) Millennium Ecosystem Assessment (MEA). (2005). In: Ecosystems and human Well-being. Island Press, Washington, DC

Medeiros E (2016) Territorial cohesion: an EU concept. European J Spatial Dev 14:30–30. https://doi.org/10.5281/zenodo.5141339

Meerow S, Newell JP (2017) Spatial planning for multifunctional green infrastructure: growing resilience in Detroit. Landsc Urban Plan 159:62–75. https://doi.org/10.1016/j.landurbplan.2016.10.005

MIBACT (2017) Piano Strategico Di Sviluppo Del Turismo. MIBACT, Rome

Momm-Schult SI, Piper J, Denaldi R et al (2013) Integration of urban and environmental policies in the metropolitan area of São Paulo and in greater London: the value of establishing and protecting green open spaces. Int J Urban Sustain Dev 5:89–104. https://doi.org/10.1080/19463138.2013.777671

Morri E, Santolini R (2022) Ecosystem services valuation for the sustainable land use management by nature-based solution (NbS) in the common agricultural policy actions: a case study on the Foglia River basin (Marche region, Italy). Land 11:57. https://doi.org/10.3390/land11010057

Myers N, Mittermeier RA, Mittermeier CG et al (2000) Biodiversity hotspots for conservation priorities. Nature 403:853–858. https://doi.org/10.1038/35002501

Nakazawa M (2021) Fmsb: functions for medical statistics book with some demographic data

Nieto-Romero M, Oteros-Rozas E, González JA, Martín-López B (2014) Exploring the knowledge landscape of ecosystem services assessments in Mediterranean agroecosystems: insights for future research. Environ Sci Pol 37:121–133. https://doi.org/10.1016/j.envsci.2013.09.003

Nogué J, Sala P (2018) Landscape, local knowledge and democracy: the work of the landscape Observatory of Catalonia. Routledge

Noya A, Garmendia E (2012) Place-based policies and the territorial approach to development. In: OECD (ed) Promoting growth in all regions. OECD Publishing. https://doi.org/10.1787/9789264174634-en

Oksanen F, Blanchet FG, Friendly M, et al (2017) vegan: Community Ecology Package. R package version 2.4–4. https.CRANR-projectorg/package=vegan

Opdam P, Coninx I, Dewulf A et al (2015a) Framing ecosystem services: affecting behaviour of actors in collaborative landscape planning? Land Use Policy 46:223–231. https://doi.org/10.1016/j.landusepol.2015.02.008

Opdam P, Albert C, Fürst C et al (2015) Ecosystem services for connecting actors—lessons from a symposium. Change Adapt Socio-Ecol Syst 2:1. https://doi.org/10.1515/cass-2015-0001

Osti G (2006) Nuovi asceti. Consumatori, imprese e istituzioni di fronte alla crisi ambientale. Il Mulino, Bologna

Oteros-Rozas E, Martín-López B, González JA et al (2014) Socio-cultural valuation of ecosystem services in a transhumance social-ecological network. Reg Environ Chang 14:1269–1289. https://doi.org/10.1007/s10113-013-0571-y

Pazzagli R (2023) Che fine ha fatto la SNAI? Inghiottita dal PNRR. Civiltà Appennino, November 10, 2023. Available at: https://www.civiltaappennino.it/2023/11/10/che-fine-ha-fatto-la-snai-inghiottita-dal-pnrr/

Palomo I, Felipe-Lucia MR, Bennett EM et al (2016) Disentangling the pathways and effects of ecosystem service co-production. In: Advances in ecological research. Elsevier, pp 245–283

Panagos P, Borrelli P, Poesen J et al (2015) The new assessment of soil loss by water erosion in Europe. Environ Sci Pol 54:438–447. https://doi.org/10.1016/j.envsci.2015.08.012

Pascual U, Phelps J, Garmendia E et al (2014) Social equity matters in payments for ecosystem services. Bioscience 64:1027–1036. https://doi.org/10.1093/biosci/biu146

Pascual U, Balvanera P, Díaz S et al (2017) Valuing nature's contributions to people: the IPBES approach. Curr Opin Environ Sustain 26–27:7–16. https://doi.org/10.1016/j.cosust.2016.12.006

Pe'er G, Zinngrebe Y, Moreira F et al (2019) A greener path for the EU common agricultural policy. Science 365:449–451. https://doi.org/10.1126/science.aax3146

Pellizzoni L, Osti G (2003) Sociologia dell'ambiente. Il Mulino

Peng J, Wang X, Liu Y et al (2020) Urbanization impact on the supply-demand budget of ecosystem services: decoupling analysis. Ecosyst Serv 44:101139. https://doi.org/10.1016/j.ecoser.2020.101139

Petrosillo I, Zurlini G, Corlianò ME et al (2007) Tourist perception of recreational environment and management in a marine protected area. Landsc Urban Plan 79:29–37. https://doi.org/10.1016/j.landurbplan.2006.02.017

Pierantoni I, Sargolini M (2021) Protected areas and local communities. A challenge for inland development. List, Trento

Pittau MG, Zelli R, Gelman A (2010) Economic disparities and life satisfaction in European regions. Soc Indic Res 96:339–361. https://doi.org/10.1007/s11205-009-9481-2

Plieninger T, Dijks S, Oteros-Rozas E, Bieling C (2013) Assessing, mapping, and quantifying cultural ecosystem services at community level. Land Use Policy 33:118–129. https://doi.org/10.1016/j.landusepol.2012.12.013

Pueyo-Ros J (2018) The role of tourism in the ecosystem services framework. Land 7:111. https://doi.org/10.3390/land7030111

Queiroz C, Meacham M, Richter K et al (2015) Mapping bundles of ecosystem services reveals distinct types of multifunctionality within a Swedish landscape. Ambio 44(Suppl 1):S89–S101. https://doi.org/10.1007/s13280-014-0601-0

Quintas-Soriano C, García-Llorente M, Norström A et al (2019) Integrating supply and demand in ecosystem service bundles characterization across Mediterranean transformed landscapes. Landsc Ecol 34:1619–1633. https://doi.org/10.1007/s10980-019-00826-7

Raudsepp-Hearne C, Peterson GD, Bennett EM (2010) Ecosystem service bundles for analyzing tradeoffs in diverse landscapes. Proc Natl Acad Sci 107:5242–5247. https://doi.org/10.1073/pnas.0907284107

Rieb JT, Chaplin-Kramer R, Daily GC et al (2017) When, where, and how nature matters for ecosystem services: challenges for the next generation of ecosystem service models. Bioscience 67:820–833. https://doi.org/10.1093/biosci/bix075

Rockström J, Gupta J, Qin D et al (2023) Safe and just Earth system boundaries. Nature 619:102–111. https://doi.org/10.1038/s41586-023-06083-8

Romagnoli L, Mastronardi L (2020) Can local policies reduce the gap between 'centers' and 'inner areas'? The case of Italian municipalities' expenditure. Economies 8:33. https://doi.org/10.3390/economies8020033

RStudio Team (2021) RStudio: Integrated Development Environment for R

Ruckelshaus M, McKenzie E, Tallis H et al (2015) Notes from the field: lessons learned from using ecosystem service approaches to inform real-world decisions. Ecol Econ 115:11–21. https://doi.org/10.1016/j.ecolecon.2013.07.009

Saidi N, Spray C (2018) Ecosystem services bundles: challenges and opportunities for implementation and further research. Environ Res Lett 13:113001. https://doi.org/10.1088/1748-9326/aae5e0

Sala OE, Stuart Chapin F III et al (2000) Global biodiversity scenarios for the year 2100. Science 287:1770–1774. https://doi.org/10.1126/science.287.5459.1770

Salata S, Giaimo C, Alberto Barbieri C, Garnero G (2020) The utilization of ecosystem services mapping in land use planning: the experience of LIFE SAM4CP project. J Environ Plan Manag 63:523–545. https://doi.org/10.1080/09640568.2019.1598341

Sargolini M (2013) Urban landscapes: environmental networks and quality of life. Springer, Milan, New York

Sargolini M, Gambino R (2016) Mountain landscapes. List

Sargolini M, Pierantoni I, Polci V, Stimilli F (2022) Progetto Rinascita Centro Italia. In: Nuovi sentieri di sviluppo per l'Appennino Centrale interessato dal sisma del 2016. CARSA Edizioni

Sassen S (2018) Cities in a world economy. SAGE Publications

Sereni E (1961) Storia del paesaggio agrario italiano, 20° edizione. Laterza, Bari

Sharp R, Tallis H, Ricketts T et al (2014) VEST user's guide. The Natural Capital Project, Stanford, CA

SINAB (2020) Biologico in cifre

Sitas N, Prozesky H, Esler K, Reyers B (2014) Exploring the gap between ecosystem service research and management in development planning. Sustainability 6:3802–3824. https://doi.org/10.3390/su6063802

SNAI Alto Maceratese (2019) SNAI Local development strategy "La rinascita dei territori nel rapporto lento-veloce"

Sonderegger G, Oberlack C, Llopis J et al (2020) Telecoupling visualizations through a network lens: a systematic review. Ecol Soc 25:1. https://doi.org/10.5751/ES-11830-250447

Sørensen JFL (2014) Rural–Urban differences in life satisfaction: evidence from the European Union. Reg Stud 48:1451–1466. https://doi.org/10.1080/00343404.2012.753142

Spangenberg JH, Beaurepaire AL, Bergmeier E et al (2018) The LEGATO cross-disciplinary integrated ecosystem service research framework: an example of integrating research results from the analysis of global change impacts and the social, cultural and economic system dynamics of irrigated rice production. Paddy Water Environ 16:287–319. https://doi.org/10.1007/s10333-017-0628-5

Spyra M, Kleemann J, Cetin NI et al (2019) The ecosystem services concept: a new Esperanto to facilitate participatory planning processes? Landsc Ecol 34:1715–1735. https://doi.org/10.1007/s10980-018-0745-6

Stenner K (2010) The authoritarian dynamic, Illustrated edition. Cambridge University Press, New York

Taiyun W, Viliam S (2021) R package "corrplot": visualization of a correlation matrix

Tallis H, Kennedy CM, Ruckelshaus M et al (2015) Mitigation for one & all: an integrated framework for mitigation of development impacts on biodiversity and ecosystem services. Environ Impact Assess Rev 55:21–34. https://doi.org/10.1016/j.eiar.2015.06.005

TEEB (2010) The economics of ecosystems and biodiversity: mainstreaming the economics of nature: a synthesis of the approach, conclusions and recommendations of TEEB. TEEB

Teixeira H, Lillebø AI, Culhane F et al (2019) Linking biodiversity to ecosystem services supply: patterns across aquatic ecosystems. Sci Total Environ 657:517–534. https://doi.org/10.1016/j.scitotenv.2018.11.440

Territorial Agenda 2030 (2020) Territorial Agenda of the European Union 2030–a future for all places, agreed at the informal ministerial meeting of ministers Responible for spatial planning and territorial development on 01 December. Germany

Tscharntke T, Grass I, Wanger TC et al (2021) Beyond organic farming—harnessing biodiversity-friendly landscapes. Trends Ecol Evol 36:919–930. https://doi.org/10.1016/j.tree.2021.06.010

Turkelboom F, Leone M, Jacobs S et al (2018) When we cannot have it all: ecosystem services trade-offs in the context of spatial planning. Ecosyst Serv 29:566–578. https://doi.org/10.1016/j.ecoser.2017.10.011

Turner MG (1989) Landscape ecology: the effect of pattern on process. Annu Rev Ecol Syst 20:171–197. https://doi.org/10.1146/annurev.es.20.110189.001131

Turner MG, Gardner RH, O'Neill RV (2001) Landscape ecology in theory and practice: pattern and process. Springer, New York

Turner MG, Gardner RH (2015) Landscape ecology in theory and practice: pattern and process, Second edition. ed. Springer, New York

United Nations U (2016) The world's cities in 2016. United Nations

UNIVPM (2019) Valutazione e quantificazione delle emissioni in atmosfera nella regione marche. Università Politecnica delle Marche, Ancona

Vasiljević N, Radić B, Gavrilović S et al (2018) The concept of green infrastructure and urban landscape planning: a challenge for urban forestry planning in Belgrade. Serbia iForest 11:491–498. https://doi.org/10.3832/ifor2683-011

Viesti G (2016) Disparità regionali e politiche territoriali in Italia nel nuovo secolo. In: Le Regioni Europee. Politiche per la Coesione e Strategie per la Competitività. Franco Angeli, Milano

Villamagna AM, Angermeier PL, Bennett EM (2013) Capacity, pressure, demand, and flow: a conceptual framework for analyzing ecosystem service provision and delivery. Ecol Complex 15:114–121. https://doi.org/10.1016/j.ecocom.2013.07.004

WB (2009) World development report 2009, Spatial Disparities and Development Policy. World Bank, Washington

Wickham H (2016) ggplot2: elegant graphics for data analysis

Wickham H, François R, Henry L, Müller K (2021) Dplyr: a grammar of data manipulation

Wunder S, Börner J, Ezzine-de-Blas D et al (2020) Payments for environmental services: past performance and pending potentials. Annu Rev Resour Econ 12:209–234. https://doi.org/10.1146/annurev-resource-100518-094206

Yu Z, Liu X, Zhang J et al (2018) Evaluating the net value of ecosystem services to support ecological engineering: framework and a case study of the Beijing Plains afforestation project. Ecol Eng 112:148–152. https://doi.org/10.1016/j.ecoleng.2017.12.017

Zalasiewicz J, Williams M, Haywood A, Ellis M (2011) The Anthropocene: a new epoch of geological time? Philos Trans R Soc A Math Phys Eng Sci 369:835–841. https://doi.org/10.1098/rsta.2010.0339

Zhang D, Huang Q, He C et al (2019) Planning urban landscape to maintain key ecosystem services in a rapidly urbanizing area: a scenario analysis in the Beijing-Tianjin-Hebei urban agglomeration, China. Ecol Indic 96:559–571. https://doi.org/10.1016/j.ecolind.2018.09.030

Zoderer BM, Tasser E, Carver S, Tappeiner U (2019) Stakeholder perspectives on ecosystem service supply and ecosystem service demand bundles. Ecosyst Serv 37:100938. https://doi.org/10.1016/j.ecoser.2019.100938

Zulian G, Maes J, Paracchini ML (2013) Linking land cover data and crop yields for mapping and assessment of pollination Services in Europe. Land 2:472–492. https://doi.org/10.3390/land2030472

Index

MIX
Papier aus verantwortungsvollen Quellen
Paper from responsible sources
FSC® C105338